EDUCATION IN A COMPETITIVE AND GLOBALIZING WORLD

TEACHING AND LEARNING

PRINCIPLES, APPROACHES AND IMPACT ASSESSMENT

EDUCATION IN A COMPETITIVE AND GLOBALIZING WORLD

Additional books in this series can be found on Nova's website under the Series tab.

Additional e-books in this series can be found on Nova's website under the eBooks tab.

EDUCATION IN A COMPETITIVE AND GLOBALIZING WORLD

TEACHING AND LEARNING

PRINCIPLES, APPROACHES AND IMPACT ASSESSMENT

MALCOLM VARGAS
EDITOR

nova
publishers
New York

Library of Congress Cataloging-in-Publication Data

Names: Vargas, Malcolm, editor.
Title: Teaching and learning : principles, approaches and impact assessment /editor, Malcolm Vargas.
Other titles: Teaching and learning (Nova Science Publishers : 2016)
Description: Hauppauge, New York : Nova Science Publisher's, Inc., 2016. |
Series: Education in a competitive and globalizing world | Includes index.
Identifiers: LCCN 2016017092 (print) | LCCN 2016022901 (ebook) |
ISBN 9781634852289 (hardcover) | ISBN 9781634852555 (Ebook) |
ISBN 9781634852555 ()
Subjects: LCSH: Distance education. | Social work education. | Science--Study and teaching. |
Educational psychology. | Educational evaluation.
Classification: LCC LC5800 .T42 2016 (print) | LCC LC5800 (ebook) | DDC
371.102--dc23
LC record available at https://lccn.loc.gov/2016017092

Published by Nova Science Publishers, Inc. † New York

CONTENTS

PREFACE

This book discusses principles, approaches and impact assessments of teaching and learning.

Chapter 1 - Teaching and learning in higher education have undergone a paradigm shift, transgressing from traditional face-to-face classroom to that of distance learning. The notion of teaching and learning via distance involves a number of means, namely text-based unit and course materials, satellite broadcasting, video conferencing, and more recently online learning. These teaching and learning modes, flexible and convenient, offer individuals more flexibility, freedom, and autonomy. Achievements of quality learning are more attractive, enabling part-time and full-time workers to strive for formal qualifications.

The notion of online teaching and learning, in particular, is of significance as it enables the linking of people and cultures, despite the separation of time and space. This pedagogical practice acknowledged worldwide by institutions for its fast-paced, insightful delivery and innovations. Having said this, however, the authors also recognize that online learning does pose a number of critical issues and problems for consideration, for example: (i) how do the authors assist learners to achieve their optimal bests?, and (ii) how do the authors instill student motivation, especially when face-to-face engagement is predominantly non-existence?

In alignment with these pervasive questions, educators and researchers have focused on different areas of research development that may relate to the potency of distance education, via technological pathways. One area of inquiry that may have educational implications for consideration is *proactive engagement* that may assist in the achievement of optimal best. How do the authors scaffold and encourage learners to actively engage in the learning processes with a view to succeed, academically? What preventive measures could the authors put in place to deter deficit and negative practices (e.g., work-avoidance behaviors)?

In this chapter, the authors explore the importance of *distance education*, focusing on a number of contemporary issues that require continuing conceptualizations and research development. The authors offer and explore a number of imperative issues, namely: (i) insightful focus on strategies, pathways, and theoretical orientations that may act, in tandem, to facilitate achievement bests and enriched personal experiences (e.g., the use of technological advances for student scaffolding), and (ii) consideration of recommendations for future research development (e.g., an inquiry into the self-efficacy of online learning). The authors postulate that their discussion would provide insightful dialogues for ongoing theorizations and conceptualizations of distance education.

Chapter 2 - In education, the responsibility is not for something but towards someone. In the actual context of the delicate process of teaching, it is no longer sufficient to transfer knowledge and competences. The challenge is to create a space in which students can train in and experiment with personal and relational abilities, gaining self-awareness of their resources and limits.

Overtaking traditional ideas about learning processes, the relational approach encourages educators to think of education in an innovative way. It is clear from the literature that the accuracy of a scientific approach to social work education is certainly important, but there is a range of different ways in which students can be engaged in learning. In relational social work education, teaching and learning become a simultaneous process focused on social work principles, such as empowerment, peer facilitation, reflexivity, auto-evaluation, valorization and involvement of service users, and emotional elaboration. Innovative activities and group work stimulate the active engagement of students, permitting them to discover and work on their own personal motivations for becoming social workers. If "relational" means that both actors are in a truly mutual relationship, aiding each other, it is more appropriate to speak about a facilitator's role than a teacher's role and about group encounters rather than lessons. The idea is to establish a reciprocal exchange of knowledge, experiences, resources, emotions, and life stories between different actors, such as professors, service users, and students. In recent years, several research projects have been conducted for the purpose of collecting information about and better understanding the effects of this approach to teaching and learning.

Chapter 3 - Education and teaching-learning process are living concepts that have been revisited in the last years. Nowadays, educators (formal, non-formal and informal ones) have included several factors that were traditionally excluded from the classical vision of the education process. That is the case of emotions and affective dimension.

This chapter presents the results of some recent works where the affective dimension is involved in order to enhance the effectiveness of the teaching-learning process, particularly inside the specific area of science education.

Two are the main branches of the current work. Firstly, the emotional performance of different teaching activities in Secondary Education is analysed, taking into account some practices are student-centered and others are traditional ones. Data collected evidenced that the first ones drove to a better emotional and academic performance, whereas the second ones clearly kept the well-known drawbacks and difficulties of traditional science education.

Secondly, the emotional performance of university students was analysed and correlated with the general tendency of the teacher. Again, a constructivist methodology seems to be more efficient in terms of promoting positive emotions.

It is more than evident that the ways forward an excellent teaching activity cannot disclaim the affective considerations of students. Otherwise the learning process is doomed to failure.

Chapter 4 - Peer-assessment is an educational arrangement in which students are actively involved in assessing their peers' work; it can be perceived also as a learning activity due to the benefits which arise from its enactment. In this study, the authors sought to examine the heuristics and strategies of undergraduate student teachers in the roles of peer-assessor and peer-assessee in a web-based peer-assessment context as part of a science course. The authors also aimed to examine whether the strategies that the authors were able to identify associate with students' learning gains and also with the type of feedback produced and received, respectively, in the two roles. Participants were 24 undergraduate student teachers in Primary Education from a university in Cyprus, who were studying a web-based unit on marine Ecosystems. Students worked in pairs, but on different computers, while enacting one inquiry cycle. Each step of this cycle led to the development of a learning artefact, which was uploaded by every student individually in their own web-portfolio; in total each student developed eight learning artefacts. By the end of the teaching unit, students reciprocally and individually assessed the web-portfolio of a peer-group, with the assistance of

3-Likert scale rubrics with pre-defined criteria. Students' web-portfolios were also assessed by the teacher using the same rubrics. The students were then allowed to revise their web-portfolios based on both the peer- and teacher-feedback received. Their actions on their screens were recorded during the whole intervention with the use of the River Past Software Pro. Data were collected from five sources; namely screen-video captured data, peer and teacher-feedback, pre- and post- instructional tests, pre- and post- instructional interviews and students' initial and revised learning products. Data were exposed to qualitative analysis and then also treated quantitatively through the use of non-parametric tests. Findings revealed two main assessor and three assessee profiles. In particular, assessors were distinguished into *autonomous* (profile 1) and *informed* (profile 2) and assessees into *teacher trusting* (profile 1), *teacher trusting and skeptical towards peer feedback* (profile 2) and *teacher and peer trusting* (profile 3). Overall, students' learning gains were not associated with their actions on either the peer-assessor or peer-assessee role; all students had positive learning gains at the end of the intervention, regardless of the strategies they had followed. However, assessors' profiles had statistically significant differences in the explicit guidance offered as part of the feedback produced; informed assessors offered more guidance compared to the guidance offered by autonomous assessors. In addition, assessees' strategies were found to be associated with the structural components of peer and teacher-feedback received. Time pressure was a critical factor that affected assessees' actions. The findings have implications for teaching practice as well as policy for teaching and learning in web-based learning environments.

Chapter 5 - Students commonly hold misconceptions in learning algebra. Because algebra is generally regarded as a gatekeeper for advanced mathematical studies in both secondary school and university and it provides essential training for engineers, scientists, and mathematicians, identifying and rectifying these misconceptions is crucial during the learning process. This chapter describes and defines students' misconceptions in learning mathematics in general and discusses the behaviorist and constructivist views on learning misconceptions. Many errors and misconceptions of students related to algebraic manipulations were found to result from overgeneralization and overspecialization. This chapter specifically elucidates student algebraic errors and misconceptions in variables, expressions, equations, and word problems. It further illustrates through examples that diagnostic teaching, which greatly emphasizes students' misconceptions and the role of cognitive conflict in the teaching and learning process, can be used as an effective teaching intervention to promote conceptual changes in learning algebra.

Chapter 6 - Introduction. In the teaching-learning process at university, achievement emotions prove to be very important and are attracting much attention as an important student variable. *Regulatory Teaching* is a characteristic of effective teaching that is related to students' achievement emotions. *Test anxiety* is a type of negative emotion, present in study and exam situations. *Coping strategies* is a meta-emotional variable defined as behavioral and cognitive efforts that are carried out in order to deal with stress, whereby the authors can address both external and internal stress-generating demands. The *objective* of the present research is to establish how different levels of Regulatory Teaching produce differences in test anxiety and the types of coping strategies used by university students.

Method. A total of 202 students from the University of Almeria (Spain) participated in an assessment of the teaching-learning process and achievement emotions. The following measures were used: (1) *Regulatory teaching* with the IATLP, in its Spanish validated version; (2) *Test anxiety*, with the TAI-80, in the Spanish validated version, and (3) *Coping strategies*, with the Spanish Coping Strategies Scales, also in a validated version. The analyses undertaken were cluster analyses (low-medium-high), ANOVAs and MANOVAs, with an ex post-facto design.

Results. The level of Regulatory Teaching had an effect on level of test anxiety, and coping strategies. In this study, the university students who perceived a *high level of Regulatory* Teaching had a low level of test anxiety, a low level of emotion-focused coping strategies, and a high level of problem-focused strategies. However, students who perceived a *low level of Regulatory* Teaching had a high level of test anxiety, high level of emotion-focused coping strategies, and a low level of problem-focused strategies.

Conclusion. Evidence is offered to defend that *Regulatory Teaching* is an important characteristic of effective teaching, because: (1) it protects against strong negative emotional reactions; (2) it is associated with *perceived predictability* in the academic context; (3) consequently, the student focuses on the problem, and does not need to focus on negative achievement emotions. With this result, the authors can focus intervention strategies on the design and development of regulatory teaching, in the university academic context.

Chapter 7 - The traditional mathematics courses are dominated by routine exercises that require memory skills and, at best, they appear mixed with a few modeling problems which are always direct problems. However, real life problems are quite the opposite of traditional courses exercises: they are frequently inverse problems and also tend to require mathematical modeling skills. These problems will be called *inverse modeling problems*; they were studied in previous Mathematics Education papers, and in Applied Mathematics book chapters. In this chapter, inverse modeling will be analyzed first from a theoretical viewpoint followed by several detailed examples. During the last two decades, some of these problems were widely used at the University of the Republic of Uruguay (UdelaR), in engineering courses. Recently, several teacher training courses and workshops – in Argentina, Guatemala and Mexico – focused on Calculus inverse modeling problems were conducted with prospective and in-service teachers. The results of these experiences will be analyzed and finally, conclusions about the potential and richness of inverse modeling problems in Mathematical Education will be presented.

Chapter 8 - Research studies on feedback and technology in learning evidence that technology can play an important role in facilitating feedback; however, it is still not clear how this is done nor is there agreement on what the most pressing issues for research in this area are. In order to address this gap, this paper aims to analyse the research on formative feedback in online or blended environments carried out between 2008 and 2012. A threefold analysis was performed, corresponding to three questions: what theoretical assumptions have these research studies adopted? What objectives and problematics were addressed? And lastly, what methodologies were used? The answers to these questions will provide a general overview of research in this area as well as identify trends that may guide the authors' decision-making in terms of setting priorities for future research and improving feedback practices in environments requiring the use of technology.

In: Teaching and Learning
Editor: Malcolm Vargas
ISBN: 978-1-63485-228-9

Chapter 1

ACHIEVING OPTIMAL BESTS, A FOCUS FOR DISTANCE EDUCATION

Huy P. Phan
University of New England, Armidale, NSW, Australia

ABSTRACT

Teaching and learning in higher education have undergone a paradigm shift, transgressing from traditional face-to-face classroom to that of distance learning. The notion of teaching and learning via distance involves a number of means, namely text-based unit and course materials, satellite broadcasting, video conferencing, and more recently online learning. These teaching and learning modes, flexible and convenient, offer individuals more flexibility, freedom, and autonomy. Achievements of quality learning are more attractive, enabling part-time and full-time workers to strive for formal qualifications.

The notion of online teaching and learning, in particular, is of significance as it enables the linking of people and cultures, despite the separation of time and space. This pedagogical practice acknowledged worldwide by institutions for its fast-paced, insightful delivery and innovations. Having said this, however, we also recognize that online learning does pose a number of critical issues and problems for consideration, for example: (i) how do we assist learners to achieve their optimal bests?, and (ii) how do we instill student motivation, especially when face-to-face engagement is predominantly non-existence?

In alignment with these pervasive questions, educators and researchers have focused on different areas of research development that may relate to the potency of distance education, via technological pathways. One area of inquiry that may have educational implications for consideration is *proactive engagement* that may assist in the achievement of optimal best. How do we scaffold and encourage learners to actively engage in the learning processes with a view to succeed, academically? What preventive measures could we put in place to deter deficit and negative practices (e.g., work-avoidance behaviors)?

In this chapter, we explore the importance of *distance education*, focusing on a number of contemporary issues that require continuing conceptualizations and research development. We offer and explore a number of imperative issues, namely: (i) insightful focus on strategies, pathways, and theoretical orientations that may act, in tandem, to facilitate achievement bests and enriched personal experiences (e.g., the use of technological advances for student scaffolding), and (ii) consideration of recommendations for future research development (e.g., an inquiry into the self-efficacy of online learning). We postulate that our discussion would provide insightful dialogues for ongoing theorizations and conceptualizations of distance education.

Keywords: achievement bests, zone of optimization, human optimization, successful learning, distance education, teaching strategies

INTRODUCTION

Distance education has emerged as an alternative, but effective medium that enables higher education institutions worldwide to offer courses and degree programs to mature-aged and part-time students. Embracing this medium of instruction, tertiary institutions are using various modes to communicate with their students, namely satellite video broadcasting, teleconferencing, text-based course materials and, more recently, online. Coupling with this continuous trend, there has also been a plethora of research and writings in this area where aspects that have been addressed include the benefits (Chappell, Rhodes, Solomon, Tennant and Yates, 2003; Kinuthia and Dagada, 2008; LaPointe and Reisetter, 2008; Perreault, Waldman, Alexander and Zhao, 2008) of distance teaching and learning, and strategies and programs that may be used to enhance effective learning via distance (Black, 2005; Chen and Bonk, 2008; Khare and Lam, 2008; L. M. Smith, 2006; Thomas, 2006; Zhang, Gao, Ring and Zhang, 2007). This recognition affirms the notion that distance education is becoming a norm and has important implications for instructors and learners alike.

Arising from the concept of distance education is a detail of a learning environment that is two dimensional and "non-humanistic" in nature. As a result, one important line of thinking emphasizes a need for educators to cultivate a climate that reflects the traditional classroom milieu – for example, active, exciting, interactive, and communicative. This feat has important theoretical and practical implications, especially when one considers the quest for enriched learning experiences. How does an individual achieve his/her optimal best in a subject matter when *in situ*, face-to-face scaffolding is relatively limited? How do we motivate learners of distance education to strive for academic excellence? Achievement bests and, consequently, enriched academic experiences for the distance education paradigm are not comparable with those of the traditional, face-to-face method of teaching and learning. We need to take stock of the pervasive issues that currently exist with the concept of distance education. This chapter, overall, explores in-depth the importance of distance education from the perspectives of investigative measures and research development. Our main task here is to offer a theoretical insight into major contemporary issues for consideration. By all means, as the literature indicates, there are a wide range of issues, criticisms, and recommendations offered with the collective view to improve the notion of distance education, in general. For us, in a similar vein, the focus of examination entails two main foci: (i) insightful focus on strategies, pathways, and theoretical orientations that may act, in tandem, to facilitate achievement bests and enriched personal experiences (e.g., the use of technological advances for student scaffolding), and (ii) consideration of recommendations for future research development (e.g., an inquiry into the self-efficacy of online learning).

DISTANCE EDUCATION: AN OVERVIEW

Distance education is not a new concept, but rather a form of educational discourse that has existed for more than five decades. Over the years, distance education has become a driving force that permeates the ideas of cost-effectiveness, flexibility, and convenience. The acceptance of learning by distance has resulted in the design, structuring, and implementation of different teaching tools, such as sending written materials via post, cable television, satellite, teleconferencing, and two-way video conferencing (Menchaca and Bekele, 2008). With different innovative strategies available for distance learning, the popularity of distance education on a global scale is reflected by the increase in student enrolment. Furthermore, more recent technological advances have seen an increasing trend in the utilization of teaching of units and courses online (Southernwood, 2008). In essence, the rapidly emerging technologies in developed countries have led educators to rely more on online learning as a main medium of instruction for the delivery of teaching materials (Adams, 2008; LaPointe and Reisetter, 2008; Legg, Adelman and Levitt, 2009; Perreault et al., 2008).

The various statistical figures to date suggest that online teaching and learning has increased student enrolment in various accredited degrees by more than 30% worldwide (Fletcher, 2004; Tallent-Runnels et al., 2006). In particular, as an alternative form of instruction, online learning offers mature-aged and part-time learners opportunities to develop specialized skills and to attain advanced degrees by distance education (Chappell et al., 2003; LaPointe and Reisetter, 2008). Online units and degree programs rely on interactive tools and continue to proliferate in this area of development (Dede, 2004; Fletcher, 2004; Menchaca and Bekele, 2008). Online courses, in particular, are defined as those courses that have at least 80% of the course content delivered online (Allen and Seaman, 2005). In the last couple of years, the increasing popularity of online learning has seen an emergence of full-scale digital universities, such as the African Virtual University (Juma, 2003) and the creation of remote branch campuses for international students (e.g., Monash University, Australia, has branch campuses in South Africa)(Ravjee, 2007). Furthermore, technology-inspired partnership relations established between higher education institutions and private corporations have prompted intense debates and research investigations into the effectiveness of distance learning via online pedagogical approaches. For example, the continued growth and established alliances in online learning suggest that this pedagogical approach may provide relevant cognitive tools to transform learners in an active and participated manner (Kinuthia and Dagada, 2008; Perreault et al., 2008).

The potency of distance education in higher education is reflected, in part, by the emergence and ongoing publications made in a number of well-known journals, notably *Distance education*, *Quarterly Review of Distance education*, and *Distance Learning*. The ongoing research development and publications made suggest there is a place for distance teaching and learning in higher education settings. Taking this into consideration, we posit that acceptance of the distance learning mode has important practical implications for instructors and learners alike. Specifically, one major implication, and integral to our subsequent discussion, is the creation of a climate that maybe perceived as being interactive and dynamic. Researchers and instructors (Brown, Collins and Duguid, 1989; Lave and Wenger, 1991; Northcote and Kendle, 2001) have, for example, referred to the tenets of situated cognition and enculturation (Vygotsky, 1978; Wertsch, 1985). Aligning closely to the theory of sociocultural constructivism, situated cognition suggests that learning is student-centered and learners actively participate in their environment to make meaning and understanding of the world (Almala, 2005; Khare and Lam, 2008; Southernwood, 2008).

Situated cognition is complex and indicates thoughtful consideration, such as the design, structuring, and implementation of innovative strategies that may then enhance academic performance for distance learning. Most pivotal in this examination is the emulation of a FACE-TO-FACE "classroom" milieu that may then accentuate meaningful learning in an authentic context for distance learners (Brown et al., 1989; Khare and Lam, 2008; Underhill, 2006). Furthermore Vygotsky's (1978) sociocultural framework of development suggests an inclusion of conversational dialogues and participation in a community of practice (Lave and Wenger, 1991). This theoretical paradigm poses a major challenge for educators, especially in the establishment and fostering of a learning environment that then binds learners across distances together (Correia and Davis, 2008; Salmon, 2005). A number of researchers and educators have surmised that good practice and sustainability of distance education depend greatly on the cultivation of intimate learning communities via distance. Relating closely to the principles of situated learning in a community of practice, there is a need for educators to emulate instructional policies and practices that may, in turn, strengthen learners' resolve and interest for distance learning. In particular, one practice that has been recognized as being relevant is the creation of an academic and social climate that reflects a sense of humanism and realism when one learns via distance (Woo, 2009). This absence of realism is more noticeable when learning, as is the case with some distance modes, lacks social interaction and/or dynamic communication (e.g., text-based course materials and/or teleconferencing). Collective then, effort is currently being made to project distance education as a viable option for learners worldwide. The concept of distance education is time flexible and cost-effective, offering opportunities for lifelong learning and accreditations despite the disaccord of time and space. Furthermore, as a pedagogical approach, distance education allows individuals to "attend" higher education institutions and to obtain advanced qualifications despite their work commitment or time-consumed related matters.

Having considered the significance of teaching and learning via distance, we also note that there are concerns and, more importantly, recommendations that require further examination and addressing. One serious consideration, in this case, entails the importance of achievement bests and enriched learning experiences. As indicated previously, the underlying contention that often masks the effectiveness of distance learning is the absence of instructors, *in situ* and face to face. This lack of 'real-life,' personal experience may limit learners from achieving their personal bests in different domains of functioning. Scaffolding in the forms of experiential personalization, *in situ* face-to-face interaction, and traditional pedagogical practices may, in this instance, deter and/or limit individuals from experiencing optimal achievement bests. This line of postulation is noteworthy, theoretically, and may yield some important issues for consideration and discussion. In the next section of the book chapter, we discuss the theoretical tenets of achievement bests.

THE FRAMEWORK OF ACHIEVEMENT BESTS

The study of quality learning entails a focus on personal accomplishment in a subject matter. What is the best that I have done so far? What is the best that I do for this subject matter? Our theoretical and empirical contributions in this area of student motivation and learning has involved the development of the *Achievement bests Framework* (Phan and Ngu, In press-2016; Phan, Ngu and Williams, In press-2016). The Achievement bests Framework stipulates different types of personal accomplishment for an individual in a subject matter. Refining our original focus of achievement bests (Phan et al., In press-2016), we reconceptualized this theoretical orientation to include five different types of achievements (Figure 1), namely:

i) *Historical achievement best*, defined as an individual's *past learning experience*, *knowledge*, and/or *performance* in a subject matter. For example, as shown in Figure 1, an individual's past record included his/her ability to run 100 meters in 11.75 seconds. His colleagues' records, in contrast ranged from 10.05 – 12.95 seconds. Likewise, in relation to academia, a Year 9 student, Melissa, might have only received a mark of 65/100 and ranked 56 out of 150 students in the previous year. It is plausible to contend that documentation of an individual's historical performances may inform his/her current and/or future performances.
ii) *Realistic achievement best* is defined as an individual's *indication of his/her current learning experience*, *knowledge*, and/or *performance*

in a subject matter. For example, what can an individual actually do or perform, at present? In the area of running, say, an individual may report of his ability to run 100 meters in 11.50 seconds. Likewise, in secondary school mathematics, Melissa may report on her ability to solve Algebra problems involving 'additions and subtractions' (e.g., $3x + 2x$). This achievement best component is insightful in indicating an individual's state of actual functioning.

iii) *Personal achievement best* is defined as an individual's *anticipated, but achievable positioning* regarding his/her *learning experience, knowledge*, and/or *performance* in a subject matter. Basically, the issue here relates to a realistic functioning that an individual could achieve in the course of his/her learning. For example, in an area of running, what does an individual anticipate and believe that he/she is capable of? An individual may report on his/her conviction that he/she could actually run 100 meters in 11.15 seconds. In secondary mathematics, differently, a student, Melissa, may state her conviction that she could learn and do Algebra problems pertaining to 'multiplications and divisions,' aside from 'additions and subtractions' (e.g., $x \times 2y \times 3z$). This achievement best component is informative in providing insights into an individual's state of belief of conviction in terms of his/her capability.

iv) *Optimal achievement best* is defined as an individual's *aspired positioning* regarding his/her *learning experience*, *knowledge*, and/or *performance* in a subject matter. This achievement best may or may not be possible, depending on an individual's historical account, motivational states, and personal conviction. What is an individual's state of aspiration to learn and/or to achieve in a course of action, taking into consideration his/her previous record? Referring back to our example of running, it is plausible to indicate that an individual may feel aspired to propose a personal goal for achievement: to run 100 meters in 10.75 seconds with reference to the world record of 9.58 seconds. In secondary school mathematics, likewise, Melissa, who is not that capable, may feel aspired to learn and achieve the topical theme of 'grouping symbols' (e.g., $p(p + 3)$) because several of her classmates have demonstrated that this is an achievable goal. We contend, however, that this aspired positioning for success (e.g., I aspire to learn more about the concept of 'grouping symbols of Algebra') is achievable, and reflects an individual's competence, persistence, and motivation.

v) *Ultimate achievement best* is defined as an individual's *maximum projection* of *learning*, *knowledge*, and/or *performance* in a subject matter. The notion of ultimate achievement best is similar to that of Maslow's (1962) theory of human motivation, whereby *self-actualization* is a personal need that is not easily achieved by anyone, for that matter. Some individuals, but very few perhaps, may have the inspirations, motivational beliefs, and zests to achieve such ultimate outcomes. For the majority of individuals, however, this feat of achieving 'ultimate outcomes' may not actually eventuate into any concrete form. Can an individual who has a historical account of health issues achieve such a projection of greatness (e.g., running 100 meters in 9.59 seconds)? that surpasses the world record of 9.58 seconds. Can an individual who is struggling financially become a millionaire? Likewise, a secondary school student who has difficulties in comprehending Algebra may not necessarily have the competency to solve Algebra problems that involve the topical theme of 'factorizing using common factors,' no matter how much he/she tries (e.g., factorize $x^2y + xy + xy^2$). This achievement best component, we contend, is achievable, but requires personal resolute (Phan and Ngu, 2015), inspiration, extreme competence, strong motivational beliefs and innovative instructional techniques. In essence, ultimate achievement best is a human motivating factor that strives to achieve maximum output. It is an ongoing development to reach the next level of benchmark, which should exceed the optimal best achievement.

Overall, the continuum of achievement bests may account and explain an individual's cognition, motivational beliefs, and behavioral patterns. A historical account, in this instance, may help educators to: (i) understand the achievement status of an individual in terms of his/her current competence, anticipated performance, aspiration, and future projection, and (ii) consider initiatives, pathways, and educational-social programs that could optimize ongoing achievement bests. The realistic achievement best informs the current level of achievement of an individual. Hence, it allows an individual to plan his/her personal achievement best that surpasses his/her realistic achievement best. In contrast, having achieved personal achievement best, an individual may aspire to achieve optimal best, a benchmark in which his or her peers have already achieved. The ultimate achievement best is an ongoing endeavor to achieve a benchmark, which has not yet been accomplished by anyone. We postulate that this theoretical orientation, which links past, present, and future achievement bests into one coherent trajectory, is insightful for the purpose of design and development of educational-social programs for implementation.

THE ZONE OF OPTIMIZATION

The continuum of achievement bests, as shown in Figure 1, also emphasizes the *zone of optimization*, which is defined as the 'difference' or 'range' between achievement bests, for example: Historical achievement bests ↔ realistic achievement best, realistic achievement best ↔ personal achievement best, etc. (Phan and Ngu, In press-2016). The zone of optimization is intricately linked to the *process of human optimization* (Fraillon, 2004; Phan and Ngu, 2015; Phan et al., In press-2016), which is defined *as an individual process of scaffolding and guidance that, in turn, optimize individuals to achieve their optimal best outcomes* (Phan and Ngu, In press-2016). How does an individual 'move' from his/her realistic achievement best to that of personal achievement best in a domain of functioning (e.g., running 100 meters from 11.50 seconds → 11.15 seconds)? Likewise, in school contexts, for example, how do we capitalize on personal achievement best in order to facilitate optimal achievement best?

There are a number of theoretical variables that could act as 'optimizers' in the process of human optimization, assisting individuals to reach their achievement bests (e.g., realistic achievement best → optimal achievement best). We contend that motivational processes, such as personal self-efficacy (Bandura, 1977, 1997), academic buoyancy (Martin and Marsh, 2008a, 2008b), and effective functioning (Phan, 2016; Phan and Ngu, 2015) may, for example, optimize individuals' achievement bests and subsequent learning experiences. How does theoretical tenet situate within the context of distance education? In other words, given the complexity of distance and time of distance teaching and learning, how does an instructor scaffold his/her students to reach their designated achievement bests (e.g., Student Nora wishes to reach personal achievement best in her first-year studies for the subjects Economics and Ancient History)? Aside from personal motivational processes, we argue that there are other extraneous influences that could enhance and optimize individuals' achievement bests.

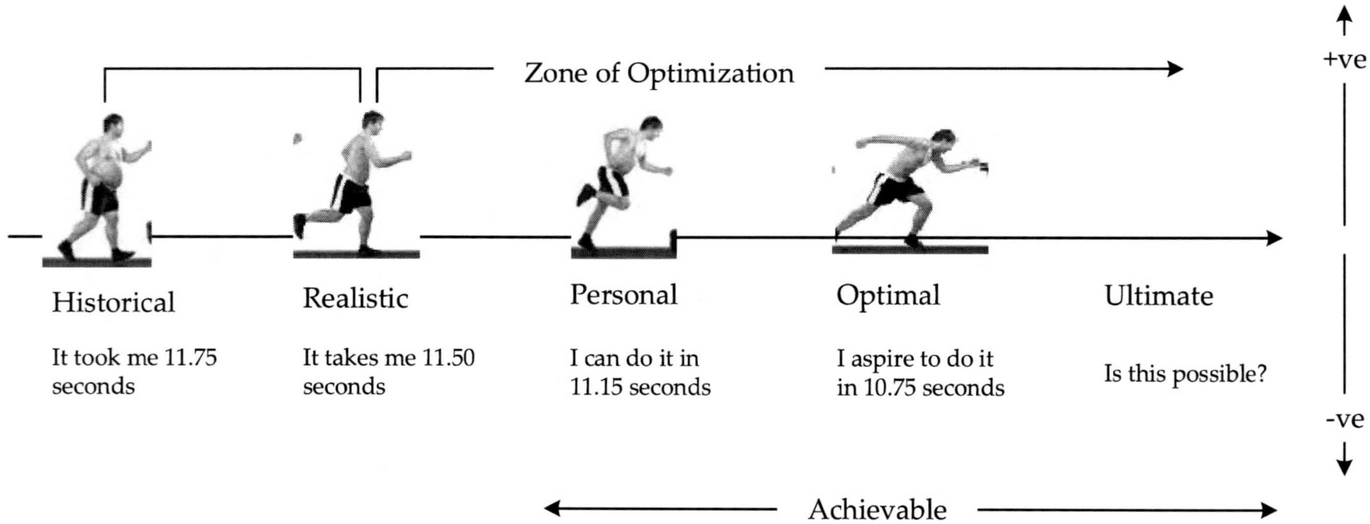

Source: Phan, H. P. and Ngu, B. H. (In press-2016). *Teaching, Learning and Psychology in the School: An Applied Approach*. South Melbourne, Victoria: Oxford University Press.

Figure 1. The Continuum of Achievement bests.

ACHIEVEMENT BESTS FROM THE PERSPECTIVE OF DISTANCE EDUCATION

The inclusion of the achievement bests framework in the context of distance education indicates some imperative issues for consideration. One major aspect, which we briefly outlined, entails initiatives for the purpose of reaching exceptional best, which varies between learners. The complexity of distance education, especially in relation to non-existing *in situ*, face-to-face interactions makes this task somewhat difficult. How does an instructor know what a first-year student's academic ambition is? What is his/her personal benchmark for achievement – for example, is it personal achievement best? Our previous discussion of achievement bests indicates a number of possibilities of achievement bests, which may depend in part on Historical achievement best. For example, a student who is 'average' in his/her academic capability (i.e., Historical achievement best) is more than likely to report moderate realistic achievement best and, possibly, personal achievement best. This student is unlikely, in this analysis, to aspire and/or to achieve anything else, for that matter.

We contend that, based on the Continuum of Achievement bests (Phan and Ngu, In press-2016; Phan et al., In press-2016), it is most advisable for learners to aspire towards optimal achievement best. Optimal achievement best, for us, is the 'reference' point for aspiration, projection, and achievement. Ultimate achievement best, in contrast, is theoretical and may not necessarily yield any concrete outcome. Personal achievement best, similarly, is achievable and but does not indicate any outstanding and/or exceptional accomplishment. Personal achievement best (e.g., I feel pretty pleased that I am able to reach the top 10 (out of 150 students) for this academic subject), in this sense, reflects a sign of contentment, but not aspiration and/or ambition. It is important for learners to exhibit an ambitious undertaking of achievement of optimal outcomes (e.g., I can go beyond my personal achievement best….). Such endeavor would serve as a source of motivation, and enable scholarly and non-scholarly excellence.

With distance education, of course, the difficulty that educators would have is their quest to scaffold students to reach optimal achievement best. Optimal achievement best in relation to distance education may encompass a number of key accomplishments, for example: (i) a student wishing to achieve an 'A' grade and rank in the top 5 for a unit of study, despite the fact he/she is studying via distance, (ii) a student wishing to graduate from his/her undergraduate degree with Cum Laude, in this case, competing with other students who may be studying on-campus, (iii) a student is able to relate his/her academic learning to real-life context and/or professional practice without any guidance, and (iv) enjoyment of proactive social relationships, via distance, utilizing innovative means. Each example, in this case, indicates aspiration towards exceptional accomplishment (e.g., achieving an excellent result), which may differentiate students from each other. There are a number of possibilities, which we discuss in this section of the chapter: instructional design, content and task delivery, situated learning, self and peer assessments, and collaborative group work. We strongly believe that these possibilities are of relevance, and align closely to the shifting paradigm found in distance education.

Our discussion, as detailed below, is for the purpose of acknowledgment, enlightenment, and consideration for further research development. One key emphasis, in particular, relates to the quest for instructors and institutions, alike, to focus on initiatives and development of policies and programs that could foster the aforementioned optimizers for achievement of optimal outcomes. The paradigm shifts in education, in general, have resulted in complex and perplexing issues for us to consider. Encouraging the achievement of optimal outcomes is a central endeavor, regardless of teaching and learning modes. From Figure 1, the 'achievable' bests include personal, optimal, and ultimate accomplishments. Historical and realistic achievement bests, as previously discussed, are informative and may serve to motivate individuals to achieve one of the three mentioned bests (e.g., personal achievement best). It is also plausible, though, for an individual to achieve beyond his/her personal achievement best. In this instance, in a progressive manner, an individual may for example move from realistic achievement best to personal achievement best, and then, optimal achievement best. We anticipate and hope that more scholarly discussions and conceptualizations could be made to address the importance of human optimization.

INSTRUCTIONAL DESIGNS

Whether distance learning via traditional course book materials or learning online, the instructional design of content materials plays a major role in the structuring and development of units and courses (Appana, 2008). One aspect of instructional design may involve the presentation, in-text, of unit materials that may stimulate intellectual curiosity and self-motivation. Learning from texts is a two-dimensional experience, where the medium of communication offers very little, if any, FACE-TO-FACE interaction. In fact, one may say that this type of learning (e.g., reading from text) is simply passive and reflects, in part, a transmission approach. The question that may be asked then is how do we motivate learners to actively participate in their learning? In a traditional tutorial class setting, instructors may use various teaching strategies to gauge students' interest and/or to stimulate their motivation for learning – for example, in an ICT tutorial class, a collaborative group assignment may require students to investigate the impact of ICT in education, or, likewise, in a biology class an inquiry-based problem approach could seek to explore the question of why humans and animals communicate (Woolfolk, 2011). These strategies may create cognitive and/or resolution, enabling students then to partake in active discussions and debates, and to come up with possible solutions. Taking this into consideration, learning by texts requires a similar approach so that there is a sense of student engagement and active participation.

Instructors in higher education institutions that still require text-based unit materials have structured and used various means to promote engagement and active participation. One key concept from our observation, for example, involves the structuring of contents in-text (i.e., text, figures, etc.). In many cases, we note that many instructional designers (e.g., Open University in United Kingdom, The University of the South Pacific (USP)) develop unit and course materials that are relatively layman in their presentations of definitions, theories, and explanations (see, for example, *ED153 School and Society, ISBN 982-03-4329-1*, USP). The structuring of contents shows a similar pattern: definitions, small portions of theories, learning activities, and revision questions for the whole module. This sequential presentation offers an effective learning approach where learning is not simply theory driven or a regurgitation of facts. Instead, there is an articulation of content delivery (i.e., lecture) and knowledge, in most cases, is actively acquired and confirmed via means of activities (e.g., Activity 3.6[1]: "Taking into consideration the fact that there are varied beliefs and assumptions, can teaching be standardized? Give reasons for your answer") and reflective thinking questions.

In the process of facilitating participation and student engagement, instructional designers have also utilized *static visuals* (e.g., figures) in their presentations. Static visualizations such as graphic organizers and concept maps have been reported to stimulate student interest, curiosity, and academic

[1] Taken from the unit, ED153, School and Society, ISBN 982-03-4329-1, USP.

achievement (Stull and Mayer, 2007). Furthermore, it has been noted that a combination of static visuals (e.g., pictures) and verbal (e.g., written text) modalities in unit presentations helps learners in their processing of information. For example, some researchers (Black, 2005; Clark and Mayer, 2008; Jitendra, 2002) have contended that visual representations (e.g., graphic organizers in text) placed with texts help strengthen the development of cognitive skills, such as evaluation and reflective and critical analysis. This pedagogical strategy of visualizations has also been included and used with online teaching and learning. Technological advances have allowed instructors to use dynamic visuals, such as animations and animated pedagogical agents (APAs) to scaffold their online teaching (Baylor and Ryu, 2003; Chang and Fisher, 2003; Welch, Blackmon, Liu, Mellers and Stark, 1996; Woo, 2009). APAs that show realistic visualizations or have human-liked characteristics (e.g., "Herman the bug" (Wouters, Paas and van Merriënboer, 2008)) are more likely to instill experiential feelings of reality and intimacy. This recognition entails the belief that animations may, but not necessarily help to create a sense of authenticity and companionship. Furthermore, the inclusion of dynamic visualizations is integral to the stimulation individuals' excitement, interest, and liking for a unit.

Overall then, from this mentioning, it is plausible to postulate that pedagogical instructions could serve to scaffold students' learning experiences in order for them to achieve optimal outcomes. Innovations, involving technologies and other pedagogical advances may play a central role in facilitating in-depth understanding and comprehension of unit contents. Shifting from conventional and traditional presentations of learning objectives and theoretical contents (e.g., a theory of a concept → reflective questions) may stimulate intellectual curiosity, personal interest, and better interconnections between topical themes, learning activities, assessment tasks, etc.

CONTENTS AND TASK DELIVERY

In unit presentations, it is prudent that clear and concise objectives and learning goals are set out in a logical manner. In particular, with reference to this emphasis, we refer to the work of John Biggs and his theoretical frameworks of constructive alignment and the SOLO taxonomy (Biggs, 1999; Biggs and Tang, 2007). Basically, in simplistic terms, the alignment teaching and learning cycle is pivotal in helping to cultivate deep learning and quality outcome. Some researchers have, for example, contended that the absence of alignment in higher education institutions has often led to discord between instructional practice and students' learning objectives (Biggs, 1987; Marton and Säljö, 1976; Phan, 2009; Schmeck, 1988). For this reason, it is important then that instructors align clear learning objectives to their teaching and assessment outcomes. This accomplishment in teaching and learning alignment, in turn, promotes deeper understanding and quality in learning for individuals.

In a similar vein, in the design and write up of unit materials, instructors often include simple written notes that are conversational, but yet instructional and deliberate in their foci. An overemphasis in definitions and explanations in written notes is redundant and only serves to overload one's cognitive processing of information. This acknowledgement arises from the theoretical principles of *redundancy* (Mayer, Heiser and Lonn, 2001; Mayer and Johnson, 2008; Moreno and Mayer, 2002) and the *cognitive load theory* (Paas, Renkl and Sweller, 2003; Sweller, 1999, 2005), and indicates a need for appropriate presentation. Considering the problem of extraneous cognitive loading, it is important that written materials have minimal overload as this excessiveness may, in turn, limit engagement in deep learning. For example, excessive and complex information often overwhelms learners, leading to feelings of disaccord and hopelessness. What this calls for then is a presentation of unit materials where written texts contain simple, relevant information without any excessive overtone. Rather than presenting complex detailed information that may overload learners, it is more prudent to integrate written texts with diagrams and figures that may scaffold and explain a particular learning concept. With reference to this idea and suggestion, we refer readers to the works of Mayer and colleagues where this strategy is used (e.g., how a water pump works).

The delivery of contents via distance may also exist in the realms of digital media and online communication. Advances made with online communication suggest there are important theoretical and practical implications for teaching and learning. One notable aspect, in particular, involves the use of dynamic and static visualizations. This pedagogical approach is grounded in the principles of multimedia learning (Mayer, 2001, 2002, 2005) and Paivio's (1986) dual coding theory and suggests that learners learn better when knowledge is imparted in dual modalities – that is, verbal and visual. Drawing from this theoretical contention, and with the technological advances available, a number of educators and researchers have recently introduced dynamic visualizations in their online teaching.

Specifically, the introduction of animations and APAs (Woo, 2009) and the embedment of video clips have provided a conducive milieu for student motivation, engagement, and learning. Researchers have noted, for example, that animations with cartoon characteristics help transform a two-dimensional learning environment into a virtual world that reflects a real-life classroom (Woo, 2009). In this analysis, cartoon images and/or animations that have lifelike characteristics (e.g., voice, body gesture, and movement) are more likely to create a sense of humanization, and to instill the presence of communal learning and social practices (Lave and Wenger, 1991; Vygotsky, 1978; Wertsch, 1985). From this evidence, we suggest the adoption of Paivio's (1986) dual coding theory in the teaching of units and courses via distance; such an approach is effective in forming a climate that strengthens learners' resolve to participate in a community of practice and to engage in communal learning. From this idea, we would like to see more empirical validation of the usage in this practice, and how inclusion of visualizations serves as study strategies that enable deep learning.

In summary, similar to that of instructional designs, we contend that subject contents and task delivery may also scaffold students in their learning processes. Difficulties pertaining to distance education, such as autonomous learning without support may be overcome by the use of appropriate pedagogical measures that take into account cognitive load imposition and redundancy and split effects. We need to have a balance in subject contents and ways of delivery that ensure efficiency, enjoyment, and effectiveness of teaching and learning. Incorporation of visual representations and limiting the scope of complexity of unit contents, for instance, may facilitate better understanding of the topic, in general. Optimal achievement best in a subject area may, consequently, arise from specialized and tailored unit designs.

SITUATED LEARNING

Situated cognition or situated learning, arising from Vygotsky's (1978) sociocultural theory of development, has been discussed as an important element in understanding learning and professional development (Lave and Wenger, 1991; Wertsch, 1985; Wertsch, del Rio and Alvarez, 1995). With reference to the understanding of distance education, the quest for educators is to situate their teaching in contextualized and sociocultural contexts. In particular, apart from active engagement, the feat of situating learning in authentic situations may contributes to the facilitation of optimal best. The difficulty, however, arises when one attempts to emulate and situate learning experiences in contexts from a distance mode perspective. Educators and instructors have addressed this anticipation by suggesting the use of conversational instructions (Vygotsky, 1978) in the writing of unit materials. By the same token, this situated learning experience is manifested in the presentation of some unit and course book materials (e.g., *ED151 Human Development, ISBN 982-03-3299-0*, USP) where hypothetical scenarios involving real life situations are used. In essence, the situated placement of learning tasks in authentic contexts plays a major role in the process of internalization of knowledge (Vygotsky, 1978). Personal experiences embedded in social and cultural settings help form schemas that, in turn, serve as a basis for learners to process information at a deep level. For example, complimentary to prescribed unit texts, effective learning of subject matters may involve activities that require learners to situate themselves in a particular context. For this scenario, instructors could pose an activity that requires learners to hypothetically position themselves in a rural school setting, where the main task is to motivate secondary school students who do not want to learn. This role-play activity, in essence, requires learners to situate their understanding of motivation outside the theory context, and to apply it to a real-life situation – for instance, what s/he would do to engage students to learn (although s/he might not have had such an experience in real life).

Apart from text-based unit materials, instructional designers have also made attempts with online approach to situate learning in context. Differing in its approach, instructors have reported and recommended the use of asynchronous text-based communication forums, collaborative activities, weblogs, and visualizations (Black, 2005; Chen and Bonk, 2008; Khare and Lam, 2008; L. M. Smith, 2006; Zhang et al., 2007). Advantageously, forums and weblogs, for example, enable learners to co-exist and to converse with others despite geographical separation. Apart from serving as a medium for conversational dialogues, these forms of communication help in the creation of virtual learning communities (Correia and Davis, 2008; Lave and Wenger, 1991; Salmon, 2005) that then transform learners, cognitively. In essence, a provision for learners to engage in electronic communications with others contributes to learning experiences relating to situated cognition. For example, messages and questions posed on forums (e.g., "What is the use of animal studies in dealing with issues such as the effects of early experiences on human development?, Source: *ED151 Human Development, ISBN 982-03-3299-0*, USP) provide opportunities for learners to discuss, debate, and reflect on their responses and answers. Here, some learners may situate their responses that are based on prior knowledge and/or personal experiences. By the same token, these situated responses are experiential in nature and reflect an authentic approach to understanding. Collaborating with this contention, some researchers have connoted the notion that asynchronous and synchronous text-based communications (e.g., wikis) establish a sense of similarity, social belongingness, and interest in learning (Black, 2005; Davidson-Shivers, Tanner and Muilenburg, 2000; Dehler and Parras-Hernandez, 1998; L. M. Smith, 2006; Thomas, 2006; Zhang et al., 2007).

Relating to the notion of situated learning, online pedagogical approach helps create a virtual classroom climate where, similar to FACE-TO-FACE classroom settings, there is a sense of connectedness (Chen and Bonk, 2008; Thomas, 2006; Trentin, 2009; Zhang, et al., 2007). In essence, the significance of online discourse lies in its capacity to provide a platform for dynamic and ongoing interactions, rather than isolation and separation between people. This consideration compells a need then, in our view, for instructors to incorporate technological advances in order to emulate a social milieu that reflects the premise of situated learning. The availability of emails, forums, podcasts, and posted messages and replies signifies an approach to learning that is communal and meaningful.

In summation, the notion of authenticity and situated cognition may an important role in the optimization of optimal achievement best. Understanding of subject contents may often require recognition of real-life relevance and/or practical application. Individuals may strive for optimal achievement best, using their previous historical track records and current knowledge as sources of information. Achieving optimal best in a subject matter may, in this instance, involve improvement of knowledge for the purpose of practical application. This line of postulation is purposive, especially for the reasons of acquiring knowledge for daily use.

SELF AND PEER ASSESSMENT AND EVALUATION

The learning-assessment alignment system is integral to the teaching and learning process (Biggs and Tang, 2007; Keppell and Carless, 2006; Sahin, 2008; H. Smith, Cooper and Lancaster, 2002), forming the basis for quality teaching and learning outcome. This milestone has been used in higher education institutions to attain academic success, and is considered as being a cornerstone for structuring learning goals and objectives. Assessment tasks used often reflect learners' cognitive thinking and various levels of understanding, taking into consideration Biggs' SOLO taxonomy. In the traditional FACE-TO-FACE classroom, there are various types of assessments – for example, group and individual oral presentations to assess and evaluate learners' understanding and communication skills. This assessment practice, similar to that of reflective journals, is effective in degree programs that require social interaction, communication, and demonstration of skills (e.g., Education). In contrast, the limitations of distance education suggest that other methods may be required to assess and evaluate student achievement and learning. For example, the nature of distance mode makes oral discourse and group presentation relatively difficult and problematic, thereby positing that other avenues of assessment may be needed. In particular, online strategies such as asynchronous and synchronous communications (e.g., Adobe Connect) allow the gauging of one's own performance and academic achievement by means of evaluative feedback from others. There is a medium available for learners to express their evaluations and opinions of others' performances; in turn, feedbacks and comments provided by others may be used as information in the process of self-reflection and monitoring (Ma, 2009; O'Grady and Choy, 2008; Slack, Beer, Armitt and Green, 2003). As a suggestion then, perhaps, instructors could allocate problem-based assessment tasks that requires the posting of answers and responses online for others to peer-review, assess, and evaluate.

Traditional unit materials also allow learners to self-evaluate and reflect their own learning experiences and performances. The nature of distance mode of learning suggests that evaluation and reflection of one's own understanding may require activities and questions that are embedded in the unit materials, themselves. One example that has often been used in Initial Teacher Education degree programs involves the development of journals and/or portfolios that demonstrate deep knowledge and understanding of units of studies. It has been noted that, in this instance, reflective journal writing and the development of portfolios assist learners to engage and to reflect in their learning (Bailey, 1990; Oscarson, 1989; Phan, 2009). The collection of individual pieces of evidence, commonly known as 'artifacts,' and the articulation of a portfolio help learners acquire relevant skills (e.g., evaluation) that may then scaffold their learning (Barrett and Carney, 2005; Jonassen, 1991).

The question for consideration then, is how does the process of assessment and evaluation optimize achievement bests from the perspective of distance education? Recognizing the central role of assessments (e.g., the use of peer assessment), in general, places emphasis on the importance of student autonomy (Deci and Ryan, 2008; Moller, Friedman and Deci, 2006).

Flexibility and opportunities to negotiate assessment tasks, for example, emphasizes the saliency of autonomous learning and independent decision making. This pedagogical practice, we contend, aligns closely to the tenets of distance learning, whereby learners are expected to demonstration self-regulation and independency. With this in mind, we believe that autonomous assessment processes, as detailed in the preceding sections, may yield achievement bests in the form of autonomous thinking and behavior.

Collaborative Group Work

The constructivist paradigm indicates a number of important tenets, one of which is the emphasis on social negotiation and collaborative group work (Driscoll, 2000). This theoretical tenet arises from the assumption that learners construct their own knowledge and make deep meaning of the world in an active manner. The social milieu is instrumental in allowing learners to seek and to socially negotiate knowledge from and with others (Vygotsky, 1978). This premise has been used successfully to structure lesson plans and programs that then result in effective learning and quality outcome. For example, based on the principles of social constructivism, medical programs across a number of higher education institutions have involved the use of problem-based teaching and learning (Hendry, Phan, Lyon and Gordon, 2002). In the context of distance education, especially with online learning educators have also used the tenets of constructivism to structure their development and teaching of units. In the area of educational psychology, for example, instructors have used learning tasks that require learners to work collaboratively towards a common goal (e.g., "In your group of five, present a PowerPoint presentation that illustrates three constructivist strategies that may be used in the teaching of algebra."). In various steps, learners may first work offline individually to research for their particular sections, before reporting back to others in their groups via various online communication means (e.g., wikis, weblogs)(Black, 2005; Chen and Bonk, 2008; Thomas, 2006; Zhang et al., 2007). Consequently, incorporating collaborative group work enables active participation and various cognitive facets, such conflict, debate, and collective resolution. We contend then, that it is prudent to integrate various teaching strategies that will ensure the presence of active participation and collaboration between learners and instructors. The use of text-based materials alone is inadequate and cannot warrant effective collaboration, dialogues, or social interaction. Furthermore, distance learning without the advent of technologies poses logistic problems for learners who wish to collaborate with others. In essence, there is a warranted needed for the use of online strategies (e.g., discussion boards and wikis) that may then result in collaboration, and the development of cognitive skills such as critical thinking, higher-order reasoning by debates, etc. (Almala, 2006). These cognitive skills (e.g., critical thinking), in turn, provide a basis for self-autonomy and, more importantly, allow learners to engage in deep meaningful learning with others (Leung and Kember, 2003).

In sum, unlike individualistic learning, collaboration is a pedagogical and learning practice that may produce both educational and non-educational benefits. Distance education, of course, makes social collaboration a relatively difficult feat to implement and/or accomplish. Successful implementation of collaboration in a distance learning context, however, makes it possible for individuals to appreciate social engagement, via different technological formats. The difficulties, in this instance, entail the disconnection of time and space, compensatory alternatives of *in situ* interactions that are comparable with face-to-face experiences, and potential student disengagement, consequently as a result of realism and personalization. Active collaboration, if successful, may enhance and optimize the achievement of personal and optimal bests, academically and non-academically.

CONCLUSION

There is no denying that higher education worldwide is undergoing a rapid paradigm shift, especially in relation to the modes of teaching and learning. With many higher education institutions experiencing some forms of restructuring and financial reformulation, the notion of continuing and accepting distance education as a major reform is neither new nor surprising. From a teaching and learning point of view, one could argue that the traditional facet-to-face pedagogical method is noteworthy and surpasses expectations. Notably, as we discussed previously, the significance of direct face-to-face teaching and learning lies mainly in the experiential feelings of humanism and social interaction between learners. The ongoing dynamics experienced by learners, arising from the tenets of constructivism (Driscoll, 2000; P. L. Smith and Ragan, 1999; Vygotsky, 1978), play a contributing role in strengthening and facilitate effective learning. This method of teaching, for example, enables the use of various study strategies (e.g., problem-based inquiry) that in turn facilitate deep meaningful learning.

From a distance learning mode perspective, we need to consider the countless limitations that define this mode of teaching and learning. Specifically, one recurring issue with distance learning that often raises debates and ongoing discussions is the absence of realism, connectedness, and socialization when one learns via distance (Peters, 2003; Reisetter and Boris, 2004; Reisetter, LaPointe and Korcuska, 2007; Tallent-Runnels et al., 2006; Woo, 2009). This recognition has not gone unnoticed and has served as a benchmark for continuing research. Despite this accepted shortcoming, many higher education institutions worldwide are accepting and embracing this distance mode of delivery as a *status quo* to the teaching and learning process. Our role then, as educators and researchers, is to design, develop, and implement a variety of innovative strategies that may, in turn, provide necessary skills for achievement bests. The emphasis, from our point of view, is concerned with the merging of subject matter and teaching approaches, and a retraining in usage of alternative pedagogies. The teacher training that prepares individuals for traditional face-to-face classroom teaching is no longer a viable norm, and serious consideration must be given to include other teaching practices. It is abundantly clear that distance learning is feasible and flexible, and offers a number of advantages (e.g., cost, time) that outweigh the traditional face-to-face classroom teaching and learning settings. This quality espouses some important requirements for educators alike, namely the strengthening and encouragement of deep learning experience, and individuals' ability to be self-regulated and self-motivated.

REFERENCES

Adams, J. (2008). Understanding the factors limiting the acceptability of online courses and degrees. *International Journal on E-Learning, 7*(4), 573-587.

Allen, I. and Seaman, J. (2005). Growing by degrees - Online education in the United States, 2005. *The Sloan Consortium.*

Almala, A. H. (2005). A constructivist conceptual framework for a quality e-learning environment. *Distance Learning, 2*(5), 9-12.

Almala, A. H. (2006). Applying the principles of constructivism to a quality e-Learning environment. *Distance Learning, 3*(1), 33-40.

Appana, S. (2008). A review of benefits and limitations of online learning in the context of the student, the instructor, and the tenured faculty. *International Journal on E-Learning, 7*(1), 5-22.

Bailey, K. (1990). The use of diary studies in teacher education programs. In J. Richards and D. Nunan (Eds.), *Second language teacher education* (pp. 215-226). New York: Cambridge University Press.

Bandura, A. (1977). Self-efficacy: Toward a unifying theory of behavioral change. *Psychol Rev, 84*(2), 191-215.

Bandura, A. (1997). *Self-efficacy: The exercise of control*. New York: W. H. Freeman and Co.

Barrett, H. and Carney, J. (2005). Conflicting paradigms and competing purposes in electronic portfolio development.

Baylor, A. L. and Ryu, J. (2003). Does the presence of image and animation enhance pedagogical agent persona? *Journal of Computing Research, 28*, 373-394.

Biggs, J. (1987). *Student approaches to learning and studying*. Melbourne, Australia: Australian Council for Educational Research.

Biggs, J. (1999). *Teaching for quality learning at university*. Buckingham: Open University Press.

Biggs, J. and Tang, C. (2007). *Teaching for quality learning at university*. Maidenhead: McGraw-Hill/Society for Research into Higher Education and Open University Press.

Black, A. (2005). The use of asynchronous discussion: Creating a text of talk. *Contemporary Issues in Technology and Teacher, 5*(1), 5-24.

Brown, J. S., Collins, A. and Duguid, P. (1989). Situated cognition and the culture of learning. *Educational Researcher, 18*(1), 32-41.

Chang, V. and Fisher, D. (2003). The validation and application of a new learning instrument for online learning in higher education. In M. S. Khine and D. Fisher (Eds.), *Technology-rich learning environments: A future perspective* (pp. 1-20). Singapore: World Scientific Publishing.

Chappell, C., Rhodes, C., Solomon, N., Tennant, M. and Yates, L. (2003). *Reconstructing the lifelong learning: Pedagogy and identity in individual, organizational, and social change*. New York: Routledge Falmer.

Chen, W. and Bonk, C. (2008). The use of weblogs in learning and assessment in Chinese higher education: Possibilities and potential problems. *International Journal on E-Learning, 7*(1), 41-65.

Clark, R. C. and Mayer, R. E. (2008). Learning by viewing versus learning by doing: Evidence-based guidelines for principled learning environments. *Performance Improvement, 47*(9), 5-13.

Correia, A.-P. and Davis, N. (2008). Intersecting communities of practice in distance education: The program team and the online course community. *Distance education, 29*(3), 289-306.

Davidson-Shivers, G., Tanner, E. and Muilenburg, L. (2000). *Online discussion: How do students participate?* Paper presented at the American Educational Research Association, New Orleans, LA (ERIC Document Reproduction Service No. ED443410).

Deci, E. L. and Ryan, R. M. (2008). Self-determination theory: A macrotheory of human motivation, development, and health. *Canadian Psychology/Psychologie canadienne, 49*(3), 182-185.

Dede, C. (2004). Enabling distributed learning communities via emerging technologies - Part One. *THE Journal, 32*(2), 12-17. Retrieved 10th May, 2009, from http://www.thejournal.com/articles/16909.

Dehler, C. and Parras-Hernandez, L. H. (1998). Using computer-mediated communication (CMC) to promote experiential learning in graduate studies. *Educational Technology, 38*(3), 52-55.

Driscoll, M. P. (2000). *Psychology of learning for instruction*. Boston: Allyn and Bacon.

Fletcher, G. (2004). The future of e-learning. *THE Journal, 32*(2), 2-3. Retrieved 10th May, 2009, from http://www.thejournal.com/articles/16899.

Fraillon, J. (2004). *Measuring student well-being in the context of Australian schooling: Discussion Paper* (E. Ministerial Council on Education, Training and Youth Affairs Ed.). Carlton South, Victoria: The Australian Council for Research, December 2014. Ministerial Council on Education, Employment, Training and Youth Affairs.

Hendry, G. D., Phan, H., Lyon, P. M. and Gordon, J. (2002). Student evaluation of expert and non-expert problem-based tutors. *Med Teach, 24*(5), 544-549.

Jitendra, A. (2002). Teaching students math problem-solving through graphic representations. *Teaching Exceptional Children, 34*(4), 34-38.

Jonassen, D. (1991). Evaluating constructivist learning. *Educational Technology, 36*(9), 28-33.

Juma, M. N. (2003). The African Virtual University: Challenge and prospects. In M. Beebe, O. Oyyeyinka, K. M. Kouakpou and M. Rao (Eds.), *AfricaDotEdu* (pp. 206-225). New Delhi: Tata McGraw-Hill Publishing Company Limited.

Keppell, M. and Carless, D. (2006). Learning-oriented assessment: A technology-based case study. *Assessment in Education, 13*(2), 179-191.

Khare, A. and Lam, H. (2008). Assessing student achievement and progress with online examinations: Some pedagogical and technical issues. *International Journal on E-Learning, 7*(3), 383-402.

Kinuthia, W. and Dagada, R. (2008). E-learning incorporation: An exploratory study of three South African higher education institutions. *International Journal on E-Learning, 7*(4), 623-639.

LaPointe, L. and Reisetter, M. (2008). Belonging online: Students' perceptions of the value and efficacy of an online learning community. *International Journal on E-Learning, 7*(4), 641-665.

Lave, J. and Wenger, E. (1991). *Situated learning: Legitimate peripheral participation*. Cambridge, MA: Cambridge University Press.

Legg, T. J., Adelman, D. and Levitt, C. (2009). Constructivist strategies in online distance education in nursing. *Journal of Nursing Education, 48*(2), 64-69.

Leung, D. Y. P. and Kember, D. (2003). The relationship between approaches to learning and reflection upon practice. *Educational Psychology, 23*(1), 61-71.

Ma, A. W. W. (2009). A longitudinal study of the use of computer supported collaborative learning in promoting lifelong learning skills. *Issues in Informing Science and Information Technology, 6*, 65-86.

Martin, A. J. and Marsh, H. W. (2008a). Academic buoyancy: Towards an understanding of students' everyday academic resilience. *Journal of School Psychology, 46*(1), 53-83. doi:10.1016/j.jsp.2007.01.002.

Martin, A. J. and Marsh, H. W. (2008b). Workplace and academic buoyancy: Psychometric assessment and construct validity amongst school personnel and students. *Journal of Psychoeducational Assessment, 26*(2), 168 - 184.

Marton, F. and Säljö, R. (1976). On qualitative differences in learning - I: Outcome and process. *British Journal of Educational Psychology, 46*(1), 4-11.

Mayer, R. E. (2001). *Multimedia learning*. New York: Cambridge University Press.

Mayer, R. E. (2002). Cognitive theory and the design of multimedia instruction: An example of the two-way street cognition and instruction. *New Directions for Teaching and Learning, 89*, 55-71.

Mayer, R. E. (2005). Cognitive theory of multimedia learning. In R. E. Mayer (Ed.), *The Cambridge handbook of multimedia learning* (pp. 31-48). New York: Cambridge University Press.

Mayer, R. E., Heiser, J. and Lonn, S. (2001). Cognitive constraints on multimedia learning: When presenting more materials results in less understanding. *Journal of Educational Psychology, 93*, 187-198.

Mayer, R. E. and Johnson, C. I. (2008). Revisiting the redundancy principle in multimedia learning. *Journal of Educational Psychology, 100*(2), 380-386.

Menchaca, M. P. and Bekele, T. A. (2008). Learner and instructor identified success factors in distance education. *Distance education, 29*(3), 231-252.

Moller, A. C., Friedman, R. and Deci, E. L. (2006). A self-determination theory perspective on the interpersonal and intrapersonal aspects of self-esteem. In M. Kernis (Ed.), *Self-esteem issues and answers: A sourcebook of current perspectives* (pp. 188 - 194). New York: Psychology Press.

Moreno, R. and Mayer, R. E. (2002). Verbal redundancy in multimedia learning: When reading helps listening. *Journal of Educational Psychology, 94*, 156-163.

Northcote, M. and Kendle, A. (2001). Online assessment criteria action: Task design in contrasting tertiary education contexts. In P. Kommers and G. Richards (Eds.), *Proceedings of World Conference on educational multimedia, hypermedia and telecommunications 2001* (pp. 920-924). Chesapeake, VA: Association for the Advancement of Computing Education.

O'Grady, G. and Choy, J. (2008). *Assessing to foster and measure deep learning in problem-based learning.* Paper presented at the ABR and TLC Conference Proceedings, Orlando, Florida.

Oscarson, M. (1989). Self-assessment of language proficiency: Rationale and applications *Language Testing, 6*, 1-13.

Paas, F., Renkl, A. and Sweller, J. (2003). Cognitive load theory and instructional design: Recent development. *Educational Psychologist, 38*, 1-3.

Perreault, H., Waldman, L., Alexander, M. and Zhao, J. (2008). Graduate Business students' perceptions of online learning: A five year comparison. *The Delta Pi Epsilon Journal, 50*(3), 164-179.

Peters, O. (2003). Learning with new media in distance education. In M. G. Moore and W. G. Anderson (Eds.), *Handbook of distance education* (pp. 87-112). Mahwah, NJ: Lawrence Erlbaum.

Phan, H. P. (2009). Relations between goals, self-efficacy, critical thinking and deep processing strategies: A path analysis. *Educational Psychology, 29*(7), 777-799.

Phan, H. P. (2016). Longitudinal examination of personal well-being: Validating newly developed concepts. In E. Ortega (Ed.), *Academic achievement: Student attitudes, social influences and gender differences* (pp. 113-130). New York, NY: Nova Publisher.

Phan, H. P. and Ngu, B. H. (2015). Introducing the concept of optimized functioning: Establishing evidence for further consideration. *The International Journal of Pedagogy and Curriculum, 22*(4), 1-19.

Phan, H. P. and Ngu, B. H. (In press-2016). *Teaching, Learning and Psychology in the School: An Applied Approach.* South Melbourne, Victoria: Oxford University Press.

Phan, H. P., Ngu, B. H. and Williams, A. (In press-2016). Introducing the concept of Optimal Best: Theoretical and methodological contributions. *Education.*

Ravjee, N. (2007). The politics of e-learning in South African higher education. *International Journal of Education and Development using Information and Communication Technology, 3*(4), 27-41.

Reisetter, M. and Boris, G. (2004). What works: Student perceptions of effective elements in online learning. *Quarterly Review of Distance education, 5*(4), 277-291.

Reisetter, M., LaPointe, L. and Korcuska, J. (2007). The impact of altered realities: Implications of online delivery for learners' interactions, expectations and learning skills. *International Journal of Electronic Learning, 6*(1), 55-77.

Sahin, S. (2008). An application of peer assessment in higher education. *The Turkish Online Journal of Educational Technology, 7*(2), 5-10.

Salmon, G. (2005). *E-moderating: The key to teaching and learning online.* New York: RoutledgeFalmer.

Schmeck, R. (Ed.) (1988). *Learning strategies and learning styles.* New York, NY: Plenum.

Slack, F., Beer, M., Armitt, G. and Green, S. (2003). Assessment and learning outcomes: The evaluation of deep learning in an online course. *Journal of Information Technology Education, 2*, 305-317.

Smith, H., Cooper, A. and Lancaster, L. (2002). Improving the quality of undergraduate peer assessment: A case for student and staff development. *Innovations in Education and Teaching International, 39*(1), 71-81.

Smith, L. M. (2006). Best practices in distance education. *Distance Learning, 3*(3), 59-66.

Smith, P. L. and Ragan, T. J. (1999). *Instructional design.* New York: Wiley.

Southernwood, J. (2008). Distance learning: The future of continuing professional development. *Community Practitioner, 81*(10), 21-23.

Stull, A. T. and Mayer, R. E. (2007). Learning by doing versus learning by viewing: Three experimental comparisons of learner-generated versus author-provided graphic organizers. *Journal of Educational Psychology, 99*(4), 808-820.

Sweller, J. (1999). *Instructional design in technical areas.* Camberwell, Australia: ACER Press.

Sweller, J. (2005). Implications of cognitive load theory for multimedia learning. In R. E. Mayer (Ed.), *Cambridge handbook of multimedia learning* (pp. 19-30). New York: Cambridge University Press.

Tallent-Runnels, M. K., Thomas, J. A., Lan, W. Y., Cooper, S., Ahern, T. C., Shaw, S. M. and Liu, X. (2006). Teaching courses online: A review of the research. *Review of Educational Research, 76*(1), 93-135.

Thomas, M. J. (2006). Learning with incoherent structures: The space of online discussion forums. *Journal of Computer Assisted Learning, 18*(3), 351-366.

Underhill, A. F. (2006). Theories of learning and their implications for on-line assessment. *Turkish Online Journal of Distance education, 7*(1), 165-174.

Vygotsky, L. (1978). *Mind in society: The development of higher psychological processes.* Cambridge, MA: Harvard University Press.

Welch, R., Blackmon, T., Liu, A., Mellers, B. and Stark, L. (1996). The effects of pictorial realism, delay of visual feedback, and observer interactivity on the subjective sense of presence. *Presence: Teleoperators and Virtual Environments, 5*, 263-273.

Wertsch, J. V. (1985). *Vygotsky and the social formation of mind.* Cambridge, MA: Harvard University Press.

Wertsch, J. V., del Rio, P. and Alvarez, A. (1995). *Sociocultural studies of mind.* Cambridge: Cambridge University Press.

Woo, H. L. (2009). Designing multimedia learning environments using animated pedagogical agents: Factors and issues. *Journal of Computer Assisted Learning, 25*, 203-218.

Woolfolk, A. (2011). *Educational psychology* (11th ed.). Boston, MA: Pearson.

Wouters, P., Paas, F. and van Merriënboer, J. J. G. (2008). How to optimize learning from animated models: A review of guidelines based on cognitive load. *Review of Educational Research, 78*(3), 645-676.

Zhang, T., Gao, T., Ring, G. and Zhang, W. (2007). Using online discussion forums to assist a traditional English class. *International Journal on E-Learning, 6*(4), 623-643.

In: Teaching and Learning
Editor: Malcolm Vargas

ISBN: 978-1-63485-228-9

Chapter 2

TEACHING AND LEARNING IN SOCIAL WORK EDUCATION: THE RELATIONAL WAY

Elena Cabiati
Catholic University of Milan, Milan, Italy

ABSTRACT

In education, the responsibility is not for something but towards someone. In the actual context of the delicate process of teaching, it is no longer sufficient to transfer knowledge and competences. The challenge is to create a space in which students can train in and experiment with personal and relational abilities, gaining self-awareness of their resources and limits.

Overtaking traditional ideas about learning processes, the relational approach (Donati, 2010; Folgheraiter, 2007; Folgheraiter and Raineri, 2012) encourages educators to think of education in an innovative way. It is clear from the literature that the accuracy of a scientific approach to social work education is certainly important, but there is a range of different ways in which students can be engaged in learning (Scourfield and Taylor, 2014). In relational social work education (Cabiati, 2015), teaching and learning become a simultaneous process focused on social work principles, such as empowerment, peer facilitation, reflexivity, auto-evaluation, valorization and involvement of service users, and emotional elaboration. Innovative activities and group work stimulate the active engagement of students, permitting them to discover and work on their own personal motivations for becoming social workers. If “relational” means that both actors are in a truly mutual relationship, aiding each other (Folgheraiter, 2004), it is more appropriate to speak about a facilitator’s role than a teacher’s role and about group encounters rather than lessons. The idea is to establish a reciprocal exchange of knowledge, experiences, resources, emotions, and life stories between different actors, such as professors, service users, and students. In recent years, several research projects have been conducted for the purpose of collecting information about and better understanding the effects of this approach to teaching and learning.

Keywords: relational paradigm, social work education, reciprocity, relational guide, ongoing learning

INTRODUCTION

Currently, educating social workers is undoubtedly an important challenge. This challenge is always complicated, not only in relation to the growing complexity of social problems but also because of the ongoing evolution of social policies and the realities of the welfare system. Using the words of Trevithick (2005), social work is a highly skilled activity. In this complex context, the mission of social work education increases in its level of responsibility; to favour a participatory approach, traditional educational styles need to be revisited. Social work education must be able to design appropriate responses to unpredictable sets of circumstances, which will require skills of a different order and ‘creative’ ways of viewing practice (Eadie and Lymberry, 2007). Social workers work in difficult situations that frequently result in ethical dilemmas; however, guidelines fail to address the actual complexity of the situations in which practitioners find themselves (Banks, 2008).

Based on the idea that it is no longer sufficient to transfer knowledge and competences, the relational paradigm (Donati, 2010; Folgheraiter, 2007; Folgheraiter and Raineri, 2012) encourages learners, supporting their efforts to think and operate in an innovative way. For this reason, the relational paradigm is overtaking traditional social work education approaches.

To educate future social workers, in addition to practitioners already engaged in the field, it is beneficial “to transfer” relational social work principles at the training level (Folgheraiter, 2007; Folgheraiter e Raineri, 2012). In the relational paradigm, the modality of contributing to students’ (social-workers-to-be) education, is relational and innovative and focused on social work principles: empowerment, peer facilitation, reflection and auto-evaluation, valorization and involvement of service users, and emotional elaboration. For several years, Italian social work education has been promoting a social work education program based on the relational approach in facing social work education challenges. The theory, experiences and reflections presented in this chapter are inspired by the relational paradigm, which concerns both social work practice and social work education. This paradigm opens doors into deeper dimensions of teaching and learning.

Becoming a Social Worker: A Kaleidoscopic Responsibility in Social Work Education

In the education field, the responsibility is not for something but towards someone. This applies to all branches of education, but in social work, the theme of responsibility is especially elevated in importance. In the relational paradigm, responsibility is not thought in a formal sense but rather as an apprehension of and care for the other's vulnerability (Jonas, 1979).

It is relatively easy to affirm that education is a topic full of responsibility. It is less easy to attribute specific responsibilities to several subjects. What responsibilities are we talking about? To whom does each belong? To what extent are these responsibilities shared?

Considering the responsibility of educators in social work education, it is possible to identify several points of reflection:

- *Social work educators have a responsibility towards the profession.* Social work education is responsible for the quality of social work and its practices, for preserving the fundamental principles of social work and for promoting a positive image of the profession.
- *Social work educators have a responsibility towards the students.* Students place their professional expectations in the education system. Social work education has a responsibility to prepare practitioners to be able to work confidently over the course of time, avoiding the risk of diminishing their personal wellbeing through burnout and stress (Folgheraiter, 2014).
- *Social work educators have a responsibility towards service users.* Service users and caregivers have not only the need but also the right to receive care from competent and motivated social workers. One of

the most important purposes of education is to ensure that clients, service systems, and communities can exercise their rights to services provided by competent and skilful practitioners (IASW, 2009, p. 5).

- *Social work educators have a responsibility towards communities and societies.* This responsibility is to society generally and to the people in the communities in which they work. This includes a responsibility to work towards an inclusive society. Society demands efficiency and utility from social workers, sometimes with complex and ambivalent expectations (Altshuler and Webb, 2009).
- *Social work educators have a responsibility towards social justice.* The connection between social work and human rights has been firmly established in the literature (Healy, 2008; Ife, 2001; Reichert, 2006; Steen, 2006; Wronka, 2008). Social work education has a responsibility to promote social justice, to challenging negative discrimination and social exclusion (International code of ethics, 2004) and to stimulate practitioners to call into question unjust policies and oppressive practices. In distributing resources equitably, social workers should ensure that the resources at their disposal are distributed fairly, according to real need.
- S*ocial work educators have a responsibility towards science.* Social work education has a responsibility to promote the production and reproduction of knowledge, to develop studies and research and to maintain and elevate the status of the profession in the scientific arena.

In turn, social welfare organizations, practitioners, and societies are responsible for regulating and supporting social work education systems (Glerup and Horst, 2014). Additionally, the same students, social-workers-to-be, cannot be excused from holding personal responsibility for actively participating in and improving the social work education system.

The Failure in Educating a Student to Become a Solver

The greater the complexity of social context grows, the easier it becomes to commit the error of educating social workers to be experts with the power to solve people's life problems.

If it is true that all professions contain a balance between 'technicality' and 'indeterminacy,' this is especially important for social work educators.

The first relational social work principle is that social problems can have different solutions but can never be solved by someone. In other words, this affirms that social workers are not permitted to manipulate people's lives according to their own intentions or wishes (Folgheraiter, 2004). With regards to the education process, it is very problematic if a young student or a social worker believes that to become a social worker means to take the burden of changing other people on their own shoulders. Several authors (Prochaska et al., 1992; Senge, 1994; Seikula e Arnkil, 2006) argue that it is not that people and families are reluctant to change but that they loathe the idea of being changed by someone. A relationship created with the intention to change or better the other person or with the presumption of transforming someone in compliance with one's own ideas (Levinas, 1982) exposes the practitioner to a condition of solitude as the "solver." Furthermore, it should not be forgotten that this idea of help may raise ethical questions and encourage mistakes (Banks, 2008; Illich, 1977).

Social service organizations, media, and social expectations may also have an influence in perpetuating this misinterpretation of social work's mission. In this sense, Ayre and Calder (2010) posit that media play an important part in creating and maintaining an image of the social work profession. Meanwhile, fuelled by the media, governments and social services are perceived as holding the view that all risks to children and adults should be foreseen and manageable (Van Bijleveld et al. 2013). In this sense, the unrealistic expectations of students (who often come to the university with the need to rescue desperate people) may encourage educators to adopt a positivistic model of education focused on the resolution of people's life problems.

Conversely, the theory of non-intervention or *laissez-faire*, partly espoused by liberals, is increasingly incomprehensible and ethically unacceptable (Folgheraiter, 2012). Starting from the certainty that life difficulties are inter-subjective and do not have a standard resolution, it is possible to affirm that they cannot be solved unilaterally by expert professionals. It is time to overturn the credo always taken for granted by social work: if we exploit the cognitive energies of scientific knowledge, it is possible to eliminate people's problems (Folgheraiter, 2012). In the reality of fieldwork, even when the practice is proceduralized and the welfare system is pressed to find solutions, social workers have the need to rely on their own thinking. Regarding the teaching process, this concept encourages the surrender to transmitting positivistic models.

This process may be accomplished through explicitly teaching support for the recognition that ambiguity and uncertainty may serve as positive elements of social work practice (Roose et al., 2012; Walsh, 2006).

These reflections lead us to reject the idea that practice is the simple application of theory, relieving students from the worry of having to store theories and prescriptive models in their brains to be remembered when the time arrives to realize them in professional activity (Folgheraiter, 2006). The methodological reason for choosing the relational social work approach is to leave students free to act on their personal experiences, educating them not to solve people's life problems but to take care of other people's agency. Without denying the necessity of guaranteeing students a complete education, including theory and practice characterized by techniques, paradigms, and scientific notions, the relational paradigm has, in the last years, opened education to a new perspective.

Social work notions and competences are unquestionably important because they provide essential tools, but becoming a social worker requires abilities that cannot be enforced or stimulated only in a traditional way. These abilities, such as sensitivity, emotional control, and the capacities for team work, to be reflexive and dialogical, to come into contact with one's own life experiences, to value relational resources, to explore personal motivations and to discover themselves are resources of undeniable importance.

To Help Students Know Themselves First

From the literature, it is known that the accuracy of a scientific approach to social work education is certainly important, but there is a range of different ways in which students can engage in learning (Scourfield and Taylor, 2014). The relational paradigm in social work education suggests that it is important for students to first know themselves before focusing on other persons who need help (commonly called service users). Although social workers strive to be objective and dispassionate, it is inevitable that emotions influence their interactions and their work (Lambert and Cattani-Thompson, 1996). Joan Tronto built her theory of care on the statement that "to be a morally good person requires, amongst others, to meet the demands of caring that present themselves in his or her life" (Tronto, 1993). Relational abilities are the main skills required (Egan, 2002) and using the words of Dominelli (1998) and Lishman, (1998) "we have to know ourselves." Empathy, sensitivity, warmth, and an unconditionally positive vision (Rogers and Kinget, 1965) are identified as the relational characteristics that social workers need to have (Ferguson, 2005). To give adequate responses to others' emotions is a delicate and complex task for practitioners, but often, it is equally complicated to give a response and a name to one's own emotions (Thompson, 1991). The path of self-exploration can contribute immensely to attentiveness (Sevenhuijsen, 2014). The emotional work and investment in self is crucial to forming good social workers. With respect to this, it is very problematic to think of social workers as people who are able to keep themselves above their own emotional experience.

Recognizing emotions and speaking about them is not a simple goal for young students, but at the same time, it is a fundamental step in the learning process, necessary to prevent them from suffering from moral blindness as future social workers. Rather than trying to be there primarily for others, students can learn to recognize that they must be there for themselves first. Sevenhuijsen (2014) suggested that "by knowing ourselves, we create space for 'knowing the other,' for meeting the other-as-(s)he is, freed from projections, illusions and false expectations."

Teaching and learning approaches may encourage students to work in depth, exploring their personal ideas, worries and prejudices. This is done not to eliminate them but to know and manage them. Emphasizing the expression of positive energies and resources, the relational social work education model is guided by the purpose of promoting personal growth in students. In other words, it encourages a maturation of interpersonal skills and human qualities.

The relational social work education model aims to facilitate students' coming into contact with their own personal life experiences and gaining self-awareness of their resources and limits. If the students do not have such awareness, their own personal experiences, needs, and unconscious behaviours influence their understanding of the other person's needs (Cabiati, 2015).

Another important aim of education is to orient and support students in discovering their personal motivations for becoming social workers and engaging with them. Becoming a social worker, or an expert in human relationships, requires investing in oneself and opening up access to one's own personal resources. Working on personal motivations for becoming a social worker may help students avoid the most common traps for a helping professional. In this sense, students have to engage in self-reflection to determine whether their personal motivation may be deeply rooted in salvific motives, paternalistic attitudes, self-care, or professional omniscience. Help professionals face a real risk of becoming "disabling experts" (Illich, 1977) or reinforcing false conceptions of empathy. It is a sheer impossibility to stand in the "shoes of another," and this is certainly so when we acknowledge the principle of plurality (Sevenhujisen, 2014).

The Professor as a Facilitator of Human Relationships

One of the guiding principles of the relational approach is that educators act not only to transfer knowledge or procedures but also to encourage and support students' agency. In this framework, the traditional role of a professor is revisited, broadening the perspective of the "teaching and learning" process. What does this mean? This process can be explained by setting aside the common image of the professor engaged in a unilateral lesson and instead using the terms "facilitation" or "relational guide," adopted by the relational paradigm (Folgheraiter, 2007). In the general context of education as a "facilitator of human relationships," this means acting from a desire to contribute to human flourishing. In the specific context of social work education practice, this means that educators should encourage students to be collaborators in the project of helping themselves.

This concept applies to social work practices, too, and is connected to the idea that the possibility of increasing wellbeing in critical life situations stems from relationships between motivated people with good intentions. In practice, this appears as all of the actors involved in a true helping relationship, aiding each other (Folgheraiter, 2007). It is sociality and humanity, modulated in the appropriate ethical and scientific terms – rather than a certain, intrinsically violent, technocratic approach – that legitimates all helping practices (Folgheraiter, 2011).

As a relational facilitator, the professor facilitates discussion, stimulating student expression and encouraging them to explore their feelings in depth, if necessary, starting with talking about their life experiences.

Keeping in mind the purpose of education, the facilitator interacts with the students, stimulating group dynamics and encouraging positive relationships. This academic framework serves as a space in which to develop interpersonal skills.

From the recent Italian experience promoted by the Catholic University in Milan (Cabiati, 2015; Cabiati e Raineri, 2015), it is apparent that the characteristics of this facilitator-professor should be that he/she is a young social worker who is active on the field, is skilled in relational abilities, is not directive, and knows and applies the principles of the relational paradigm learnt during his/her own experience as a student or later through dedicated learning. Based on a peer-to-peer logic, the young age of this figure may facilitate active participation and openness in students.

As part of the education process, the relational facilitator changes the classroom environment from one in which students are passive recipients of the instructor's knowledge to one in which they are active participants in their own education (Steiner et al., 1999). In this delicate task, this figure remains aware of the purpose of the educational process but accepts a certain degree of indeterminacy.

As Prosser and Trigwell (1999) clearly indicate, the nature of what a student learns is not dictated solely by what is taught or by how it is taught. A democratic environment creates the context for a variety of emotional experiences that have the potential to influence not only learning but also teaching and motivational processes.

In other words, in the relational social work education paradigm, the educator-as-facilitator takes part in the teaching and learning process with openness and confidence, not determining from the start every step and direction of the path ahead. Stimulating active engagement, the professor-as-facilitator develops, with student collaboration, the best ways to reach educational goals. In doing so, this figure shows confidence in students' creativity and personal resources. The confidence gained allows the professor-facilitator to more easily make decisions in delicate and complicated situations. Furthermore, he/she develops his/her own balance between a sense of professionalism and a useful level of involvement, succeeding in the struggle to maintain the line between care and control (Aultman et al., 2008).

In doing this, the educator demonstrates the function of the relational guide (Folgheraiter, 2007) in an authentic way. This experience, for future social workers, is very helpful for both social work education and practice. Students experiment under the facilitation of a relational guide and over the course of their education; they assume this function, most importantly, in "out of track placements" (Doel, 2009). The "teaching and learning approach" invites young students to experiment for themselves with the social work world, autonomously but not alone, searching for collaborators in local communities with whom to deliberate and act for the purpose of enhancing wellbeing. Just as in social work practice, in fieldwork, the main challenge is to co-construct the learning process as a reflective and collaborative practice led by social work principles, such as empowerment, peer-facilitation, participation, reciprocity and self-determination.

TEACHING AND LEARNING USING EXPERIENTIAL COMPETENCES

Where is the boundary between teaching and learning? A social worker knows that the nature of social problems requires a mixture of technical knowledge (which rests predominantly on practitioners' shoulders) and life knowledge (predominantly the service users' experiences) combined in an equal and mutual relationship, able to empower both parties. In this unconventional vision of social services, the principle of reciprocity guides the establishment of the ideal relationship. In social work, in addition to its obvious deontological implications, the principle of reciprocity has a crucial methodological significance; as noted above, in any helping relationship, help is only effective when the worker accepts help from the interested parties as if they themselves were workers (Folgheraiter, 2011). Given that this principle is valid both for social work practice and for education, it is possible to affirm that the relationship between teachers and learners cannot be considered in a traditional way.

In social work education, students need to learn from educators in the same way that educators need to learn from students. Affirming this, it is possible to surmise that educators learn from students how best to teach them. Reciprocity means that not only do teachers influence students but also that students influence teachers (Aultman et al. 2008). Granting a space of reciprocal influence for the learning process may be uniquely effective.

Agreeing with this and keeping in mind the mission of social work education, there is another step. In educating students to become social workers, it is necessary to seek help from service users. Considering service users as "experts by experience" in their life problems (Croft and Beresford, 1990), their involvement is seen as providing a balanced education to potential practitioners and modelling good practice for the future (Tyler, 2006; Waterson and Morris, 2005). In preparing future social workers, clients must be considered the best communicators of their own life experiences (Valutis, 2015). The rationale for such involvement is that an understanding of the everyday experiences of service users is fundamental to equipping emergent social workers with key aspects of knowledge and expertise needed to improve services (Beresford, 2000; Anghel and Ramon, 2009). The possibility of learning from the people most involved in social services enables the development of self-awareness and communication skills in young students (Skilton, 2011).

Remembering that the education of future social workers cannot be reduced to technical training, there are concrete reasons for the involvement of service users in the academic arena; future social workers can benefit from an understanding of their life experiences and histories of coping with life difficulties.

Researchers have shown that among the various benefits of service users' engagement in the education system, there is a particularly interesting effect on changes in attitudes towards clients (Corrigan and Penn, 2015; Cabiati and Raineri, 2015). Based on the idea that social workers often share the negative view of social problems commonly held by much of the population (Beresford and Croft, 2008; Penn et al. 1994; Trevithick, 2005; James et al. 2012; Scheyett and Kim, 2004), studies have focused on the positive outcomes of service users' engagement, aiming to show a decrease in individual students' negative perceptions of social problems, such as mental illness, drug addiction, and disability. The personal interaction between "experts by experience" and social work students seems to have a greater effect than didactic intervention (Corrigan et al., 2001).

The participation of service users in educational programs may contribute to positive effects in academic settings in a number of ways. In addition to the decrease of stigmatizing attitudes towards service users and their life problems, this involvement seems to illustrate, in a concrete way, the importance of social workers being strength-focused and also of believing in the potential for service users to grow and improve. Furthermore, promoting direct interpersonal contact with people who have coped with life problems may be an effective strategy for reducing dangerous, common feelings among social workers, such as anger, diffidence and discrimination.

If, as practitioners, we can involve ourselves in the experience of relating to another human being, what we ourselves gain from this involvement is that we also can develop and learn from the encounter, about ourselves and about other people (Trevithick, 2005).

In addition, developing alternative ways of conceptualizing relationships between social workers and service users can start in the context of academic courses, where service users, acting as facilitators in the classroom, can help to break down any barriers and preconceptions that students may have (Eadie and Lymbery, 2007). Both the development of student-directed learning and the active involvement of service-users-as-educators embody a 'bottom-up' approach to change (Eadie and Lymbery, 2007).

In this framework, the process of learning happens thanks to the generous contributions of service users, whose life stories are combined with different sources of knowledge, even of human matrix.

To educate following a "teaching and learning approach" means to facilitate a dialogical process involving different actors, crossing traditional boundaries and sharing the purpose of this delicate mission. The aim of social work education guides all of the actors involved in co-constructing spaces of knowledge, listening to every voice involved and considering "teaching and learning" as an ongoing process that is never complete.

The challenge for social work education is to educate in cooperation with students and service users, co-producing knowledge and stimulating personal growth. The reciprocity of helping – allowing the helpee to be an equal and to help us in return – is not an empty slogan (Ferguson and Lavallette, 2004) or something that "is better if it exists." It is a radical turning point (Folgheraiter, 2012, pp.83) because if it is absent, that help withers and dies (Beresford and Croft, 2008).

The more that educational methods are shared, the more efficient student development will be, and the more involved teachers will become.

CONCLUSION

The "Teaching and Learning" approach stimulates a debate about how education for social workers can best prepare practitioners to exercise judgement and, above all, to best provide aid and care in their practice.

Returning to the essence of education, the task is to transmit an authentic care for the other, not to understand how to assess or judge him/her but to realize that, first of all, this encounter may be an occasion of reciprocal learning.

To enable reciprocal learning, coherence in social work education models is required, as are programs promoting the development of a learning path characterized by awareness of personal limits, humility and reciprocity. Personal characteristics limit growth resulting from inevitably assymetric reciprocity in relationships (Young, 2006).

Following the idea that it is important recognize others as equipped with unique knowledge of life, the dimensions of reciprocal exchange in learning may be considered a network of infinite contributions and connections.

The binary opposition between teachers and learners may be problematic, presenting a challenge for education in opening a way to being relational educators and relational social workers.

The hopes of educators to form good social workers become more concrete in a dialogical process, in which it is required that teachers be open, that they encourage cooperation in entering into dialogue (Arnkil and Seikkula, 2006).

In the arena of education, the kaleidoscopic responsibility towards students, professionals, teachers, and citizens may find a meeting point in the idea that if responsibility is always towards someone, a human and participatory approach is required.

The reality of social work clearly shows both the need and the convenience of envisioning education as diverging from unilateral teaching: in this delicate change of perspective, the relational paradigm inspires both education and practice promoting concretely democratic, empowering and inclusive methods.

REFERENCES

Altshuler, Sandra and Webb Jennifer Reid (2009). School social work: increasing the legitimacy of the profession. *Children and Schools*, 31(4), 207–218.

Anghel, Roxana, and Ramon Shula (2009). Service users and carers involvement in social work education: Lessons from an English case study. *European Journal of Social Work*, 12, 185–199.

Arnkil, Tom and Seikkula Jakko (2006). Dialogical meetings in social network, London, Karnac books.

Aultman, Lori Price, Williams-Johnson Meca, and Schutz Paul (2009). Boundary dilemmas in teacher–student relationships: Struggling with "the line." *Teaching and Teacher Education*, 25 636–646.

Ayre, Patrick, and Calder M. (2010). The deprofessionalisation of child protection: regaining our bearings. In: P. Ayre and M. Preston-Shoot (2010). *Children's services at the crossroads: A critical evaluation of contemporary policy for practice*. Lyme Regis: Russell House.

Banks, Sarah (2008). Critical commentary: Social work ethics. *British Journal of Social Work*, 38, 1238–1249.

Beresford, Peter (2000). Service users' knowledge and social work theory: conflict or collaboration?. *British Journal of Social Work*, 30, 489–503.

Beresford, Peter, and Croft Suzy. (2008). Democratising social work - a key element of innovation: from "client" as object, to service user as producer. *The Innovation Journal: The Public Sector Innovation Journal*, 13, 5–22.

Cabiati, Elena (2015), "Teaching and learning: an exchange of knowledge in the university among students, service users and professors," *«European Journal of social work»*, pp. 1–18.

Cabiati, Elena, and Raineri M.L. (2015), "Learning from service users involvement: A research about changing stigmatizing attitudes in social work students," *«Social work education»*.

Corrigan, Patrick W. et al. (2001), Prejudice, social distance, and familiarity with mental illness, *Schizophrenia bulletin*, 27.2 (2001): 219.

Corrigan, Patrick W., and Penn D. L. (2015). Lessons from social psychology on discrediting psychiatric stigma. *Stigma and Health*, 1, 2–17.

Doel, Mark (2009), Social work placements: a traveller's guide. London, Routledge.

Dominelli, Lena (1998). 'Anti-oppressive Practice in Context,' In: R. Adams, L. Dominelli and M. Payne (eds.) *Social Work: Themes, Issues and Critical Debates*. Basingstoke: Macmillan.

Dominelli, Lena (2004). Social work: Theory and practice for a changing profession, London, Palgrave, trad.it. Il servizio sociale. Una professione che cambia, Trento, Erickson.

Donati, Pierpaolo. (2010), Relational sociology: A new paradigm for the social sciences. London: Routledge.

Eadie, Tina, and Lymberry Mark (2007). Promoting creative practice through social work education. *Social Work Education*, 26(7), pp. 670–683.

Egan, John (2002). Accelerating change. *A report by strategic forum for construction*. London: Rethinking Construction.

Ferguson, Harry (2005). Working with violence, the emotions and the psycho-social dynamics of child protection: Reflections on the Victoria Climbié case. *Social Work Education*, 24, 781–795.

Ferguson, Ian, and Lavallette, Michael (2004), 'Beyond power discourse: alienation and social work,' *British Journal of Social Work*, vol. 34, pp. 297–312.

Folgheraiter, Fabio (2004). Relational social work: toward networking and societal practices, London, Jessica Kingsley Publishers.

Folgheraiter, Fabio (2011). Fondamenti di metodologia relazionale: La logica sociale dell'aiuto, Trento, Erickson (Fundamentals of relational methodology: The aid social logic, Trento, Erickson).

Folgheraiter, Fabio (2012). The mystery of social work. Trento, Erickson.

Folgheraiter, Fabio, and Raineri, Maria Luisa (2012). A critical analysis of the social work definition according to the relational paradigm. *«International social work»,* n.4, pp. 473–87.

Folgheraiter, Fabio (2014). Non fare agli altri (Do unto others). Il benessere in una società meno ingiusta, Trento, Erickson (The well-being in a less unjust society, Trento, Erickson).

Glerup, Cecilie, and Horst Maja (2014). Mapping 'Social Responsibility' in Science. *Journal of Responsible Innovation.* 1(1):31–50.

Healy, Lynne M. (2008), Exploring the history of social work as a human rights profession. *International Social Work*, 51, 735–748.

Ife, Jim (2001). Human rights and social work: Toward rights-based practice. New York, NY: Cambridge University.

Illich, Ivan (1977). Disabling profession, London, Marion Boyars. Trad.it. Esperti di troppo: Il paradosso delle professioni disabilitanti, Trento, Erickson, 2008 (Trad.it. Experts too: The paradox of disabling professions, Trento, Erickson, 2008).

Irish Association of Social Workers (2009). Continuing professional development policy. Retrieved from http://www.iasw.ie/attachments/86a42d72-f785-4f25-9abaed7a17c23342.PDF.

James Bawo, Ohiole Omoaregba Joyce, Osemudiamen Okogbenin Esther (2012). Stigmatising attitudes towards persons with mental illness: a survey of medical students and interns from Southern Nigeria. *Mental Illness*, 4, 32-34.

Jonas, Hans (1979). Das Prinzip Verantuwortung, Frankfurt am Main, Suhrkamp (The principle Verantuwortung, Frankfurt, Suhrkamp).

Reichert, Elisabeth (2003). Social work and human rights: A foundation for policy and practice. New York, NY: Columbia University.

Lishman, Joyce (1998). Evaluation and social work practice, London, Sage.

Lambert M.J. and Cattani-Thompson, K. (1996), Current findings regarding the effectiveness of counseling: Implication for practice. *Journal of counseling and development*, 76, 601-08.

Lévinas, Emmanuel (1982). Etica e infinito, Roma, Castelvecchi (Ethics and Infinity, Rome, Castelvecchi).

Penn, David, Guynan, Kim, Daily, Tamara, Spaulding William, Garbin, Calvin, and Sullivan, Mary (1994). Dispelling the stigma of schizophrenia: What sort of information is best?, *Schizophrenia Bulletin*, 20, 567-577.

Prochaska, Johnson, DiClemente C. and Norcross J. (1992), A tailored nutrition pilot program of how people change: applications to addictive behaviors. *«American Psychologist»*, n. 47, pp. 1102–1114.

Prosser, Michael and Trigwell Keith (1999). Understanding learning and teaching: the experience in higher education. Open University Press, Buckingham.

Rogers, Carl Ransom, and Kinget Marian (1965). Psychotérapie et relations humaines [Psycotherapy and human relationships]. Nauwelaerts: Louvain.

Roose, Rudi, Roets, Griet and Bie Maria (2012), Irony and social work: In: search of the happy Sisyphus. *The British Journal of Social Work*, 42, 1592–1607.

Scourfield, Jonathan, and Taylor Amanda (2014). Using a book group to facilitate student learning about social work. *Social Work Education* 33 (4), pp. 533–538.

Sevenhuijsen, Selma (2014). Care and attention Paper for the conference 'A meaningful life in a just society. Investigating Wellbeing and Democratic Caring.' Panel 'Rendering Care the Meaning of Politics.' Utrecht, Universiteit voor Humanistiek.

Senge, Peter et al. (1994), The fifth discipline fieldbook strategies and tools for building a learning organisation, New York, Curreny Doubleday.

Skilton, Christine Jane (2011). Involving Experts by Experience in assessing students' readiness to practise: the value of experiential learning in student reflection and preparation for practice, *Social Work Education*, 30, 299-311.

Steen, Julie (2006). The roots of human rights advocacy and a call to action. *Social Work*, 51, 101–105.

Steiner T., Stromwall L., Brzuzy S, and Gerdes K. (1999), Using Cooperative Learning Strategies in Social Work Education, *Journal of Social Work Education*, 35:2, 253–264.

Sue Steiner, Layne K. Stromwall, Stephanie Brzuzy and Karen Gerdes (1999) Using Cooperative Learning Strategies in Social Work Education, *Journal of Social Work Education*, 35:2, 253-264.

Thompson, Neil (1991). Crisis intervention revisited. Birmingham: Pepar.

Trevithick, Pamela (2005). Social work skills. Berkshire, Open University Press.

Tronto, Joan (1993). Moral Boundaries. A political argument for an ethic of care. London and New York: Routledge.

Tyler, Gina (2006). Addressing Barriers to Participation: Service User Involvement in Social Work Training. *Social Work Education*, 25, 385–392.

Valutis, Stephanie (2015). The Relationship Between Tolerance ofa mbiguity and Stereotyping: Implications for BSW Education, *Journal of Teaching in Social Work*, 35:5, 513-528.

Van Bijleveld, Ganna, Dedding Christine W. M. and Bunders-Aelen Joske F. G. (2013). Children's and young people's participation within child welfare and child protection services: a state-of-the-art review. *Child and Family Social Work*, 20, 129–138.

Walsh, Trish (2006). Two sides of the same coin: Ambiguity and complexity in child protection work. *Journal of Systemic Therapies*, 25, 38–49.

Waterson, Jan, and Morris Kate (2005). Training in 'social' work: Exploring issues of involving users in teaching on social work degree programmes. *«Social Work Education»*, n. 24, pp. 653–675.

Wronka, Joseph (2008). Human rights and social justice: Social action and service for the helping and health professions. Thousand Oaks, CA: Sage.

Young, Iria M. (2006), Asimmetrical reciprocity: on moral respect, wonder and enlarged thought, Blackwell Publishers.

In: Teaching and Learning
Editor: Malcolm Vargas
ISBN: 978-1-63485-228-9

Chapter 3

EMOTIONAL IMPLICATIONS OF SCIENCE EDUCATION: A NEW CHALLENGE FOR EXCELLENT TEACHERS

J. Sánchez-Martín**, García J. Álvarez-Gragera, J. M. Herrera-Muñoz and M. A. Dávila-Acedo

Universidad de Extremadura, Department of Science and Mathematics Education, Spain

ABSTRACT

Education and teaching-learning process are living concepts that have been revisited in the last years. Nowadays, educators (formal, non-formal and informal ones) have included several factors that were traditionally excluded from the classical vision of the education process. That is the case of emotions and affective dimension.

This chapter presents the results of some recent works where the affective dimension is involved in order to enhance the effectiveness of the teaching-learning process, particularly inside the specific area of science education.

Two are the main branches of the current work. Firstly, the emotional performance of different teaching activities in Secondary Education is analysed, taking into account some practices are student-centered and others are traditional ones. Data collected evidenced that the first ones

* Corresponding author: Email: jsanmar@unex.es.

drove to a better emotional and academic performance, whereas the second ones clearly kept the well-known drawbacks and difficulties of traditional science education.

Secondly, the emotional performance of university students was analysed and correlated with the general tendency of the teacher. Again, a constructivist methodology seems to be more efficient in terms of promoting positive emotions.

It is more than evident that the ways forward an excellent teaching activity cannot disclaim the affective considerations of students. Otherwise the learning process is doomed to failure.

Keywords: science education, emotional education, emotional performance, technology, higher education

1. INTRODUCTION

Rational mechanisms are undoubtedly in the core of our behavior, but emotional ones play a vital role that is increasingly taken into account by experts in teaching-learning process. Despite their evident relevance (one remembers with difficult the contents that studied, but it is easy to remember how one felt when studying such topic), the study of emotions and feelings was seen as being far removed from education for many years, and even more so from the teaching and learning of specific school content (Brígido et al., 2013a; Álvarez et al., 2013). Technology contents are even more distant from these considerations that are recently inserted in the science education corpus (Black and Atkin, 2005).

However, as Hargreaves (1998) puts it, *emotions are at the heart of education*. Emotions and feelings must be included in the set of variables that should be handled by the teacher for improving the development of learning and it is today recognized that the cognitive configures the affective, and vice versa, and the idea of teaching and learning as an emotional practice in which cognitive and affective processes take part is fully accepted by researchers and educators (Brigido et al., 2013b; Shapiro, 2010).

Otero (2006) notes that there is no human action without an emotion that substantiates it and makes it possible. Emotion is central to decision making (Damasio, 2010). For pupils, decision making becomes especially important when, at the end of their compulsory education, they have to decide on the direction of their future studies (Lavonen et al., 2008). Numerous studies have found that the attitudinal and emotional factors involved in learning science become increasingly negative with age, especially during secondary school (Beauchamp and Parkinson, 2008; Marba-Tallada and Márquez, 2010; Pérez and De Pro, 2013; Vázquez and Manassero, 2011). Consequently, it is not unsafe to affirm that the decline in vocational careers in science and technology in many countries may be related to the adverse emotional context surrounding science and technology learning at school (Vázquez and Manassero., 2007).

From theoretical framework of conceptual change and pupils' alternative ideas, Printrich et al. (1993) questionned *cold change*, and argued for the importance of motivation and the emotions as determinants in learning. Positive emotional states favour learning, while negative emotional states severely limit the ability to learn (Vazquez and Manassero, 2007). Conceptual change is therefore both cognitive and affective (Thagard, 2009) and teachers who ignore the affective aspects of learning may limit their pupils' conceptual change (Duit et al., 2008).

There are a variety of taxonomies for the affective domain, emotions, and feelings. For Dos Santos and Mortimer (2003), the affective is more general, and subsumes emotions, feelings, and moods. Many authors concur that emotions are the organism's automatic responses to external stimuli, whereas feelings are more permanent. In the present study, we shall be referring primarily to the emotions, although sometimes we shall refer generically to the affective dimension. Of the many definitions of emotions, we adopt that put forward by Bisquerra (2001):

> Emotions are reactions to the information we receive in our relationships in the environment. The intensity of the reaction depends on subjective assessments that we make of how this information will affect our well-being. These subjective assessments will involve prior knowledge, beliefs, personal objectives, perception of a challenging environment, etc. An emotion depends on what is important for us.

Despite this general definition, there are different conceptions of emotions from one culture to another, and even these conceptions can change over time within the same culture, so that the existence of various alternative classifications may well be justifiable. One of the most widely used classifications is that which distinguishes between basic or primary emotions and complex or secondary emotions (Francisco et al., 2005). Damasio (2010) classifies emotions as primary or basic, background, and social. According to Casacuberta (2000), basic emotions can be classified into six groups: surprise, happyness, fear, rejection, wrath and sadness. But the classification which comes closest to the nature of the current study is that which distinguishes between positive, negative and neutral emotions (Bisquera et al., 2009; Brigido et al., 2013b; Fernández-Abascal et al., 2001).

According to Manassero (2013), emotions must be included in the attributional models in order to link motivation, emotion and academic performance. It is clear and known that each student feels in a different way if he/she perceives success or failure. Positive emotions will increase the academic efforts and the self-confidence, while no success appreciation would lead him/her to scholar drop (Weiner, 1986). It is remarkable that there is no totally\emph{negative} emotions since anxiety could be a motivational force, but it could also block the student, even within adults (Goleman, 1996). Hence, a student can be blocked if facing emotional situations such as frustration, wrath or impotency inside or outside the scholar class. As (Vazquez and Manassero, 2007) pointed out, positive emotional states favour learning, while negative emotional states severely limit the ability to learn.

The relevance of emotions as activators of learning process has been poorly taken into account and almost no academic literature is found in the field of Technology education in Secondary School (Hill and Smith, 1998). Previous research indicated the need for analysing the emotions by distinguishing the different subjects of science and technology Hoeven et al., 2011; Vázquez and Manassero, 2007). Other studies show that secondary pupils have positive attitudes and emotions towards Biology and Geology and negative ones towards Physics and Chemistry (Borrachero et al., 2011; Brígido et al., 2013; Marbá-Tallada and Márquez, 2010).

Wu and Huang (2007) investigated ninth graders' cognitive, emotional, and behavioral engagement in teacher-centered and student-centered technology-enhanced classrooms. The results showed that students in the student-centered class presented significantly higher emotional engagement, and interacted more through discussion groups, but this had no impact on students' learning achievement.

In order to improve the emotional performance of science teaching, this chapter consists of two linked parts:

a) Firstly, an evaluation of the emotional implications towards a classic science classroom is given, evaluated inside university students.
b) Secondly, in an attempt of increasing the emotional performance of science teaching, several activities focused on an innovative

perspective (Multiple Intelligences Theory) are implemented at Compulsory Secondary Education.

2. The Emotions at University Science Classroom

Many previous studies have pointed out the interest and the relevance of focusing the emotional performance of students, including university ones. However, perhaps the specific evaluation of this emotional scoring on each scientific matter and its particular consideration inside the global success of the teacher's work is still unattended. For overcoming this challenge, this chapter presents some results of the emotional evaluation of a university classroom. The scientific matter was called 'Matter and energy Education' and it is included in the syllabus of preservice primary teachers, 2nd year. It is a scientific-content matter (in opposition to a didactic-content ones) that is compulsary for every student, whatever the specialization path they have.

2.1. Sample Description

This research was made on up to 86 students who were asked about the emotions felt on 'Matter and Energy Education' the year before. Sociologically, the sample is defined by the following conditions (Figure 1):

a) More than 60% are female (Subfigure 1.1).
b) Almost 90% of the sample are between 20 and 23 years old (Subfigure 1.2).
c) The large majority of the sample succeed the matter with marks between 6 and 8 (Subfigure 1.3).

As Figure clearly depicts, no exceptional circumstances are observed in this sample and the sociological description of such researching group is not out of the standard subjects at Faculty of Education.

2.2. Methodology

The data collection was carried out by means of an *ad hoc* questionnaire that was built up following Arribas (2004). Specifically, previous works such as Brígido et al. (2009), Fernández et al. (2013) and Costillo et al. (2013) pointed out some guidance lines for measuring emotional performance in these kinds of studies.

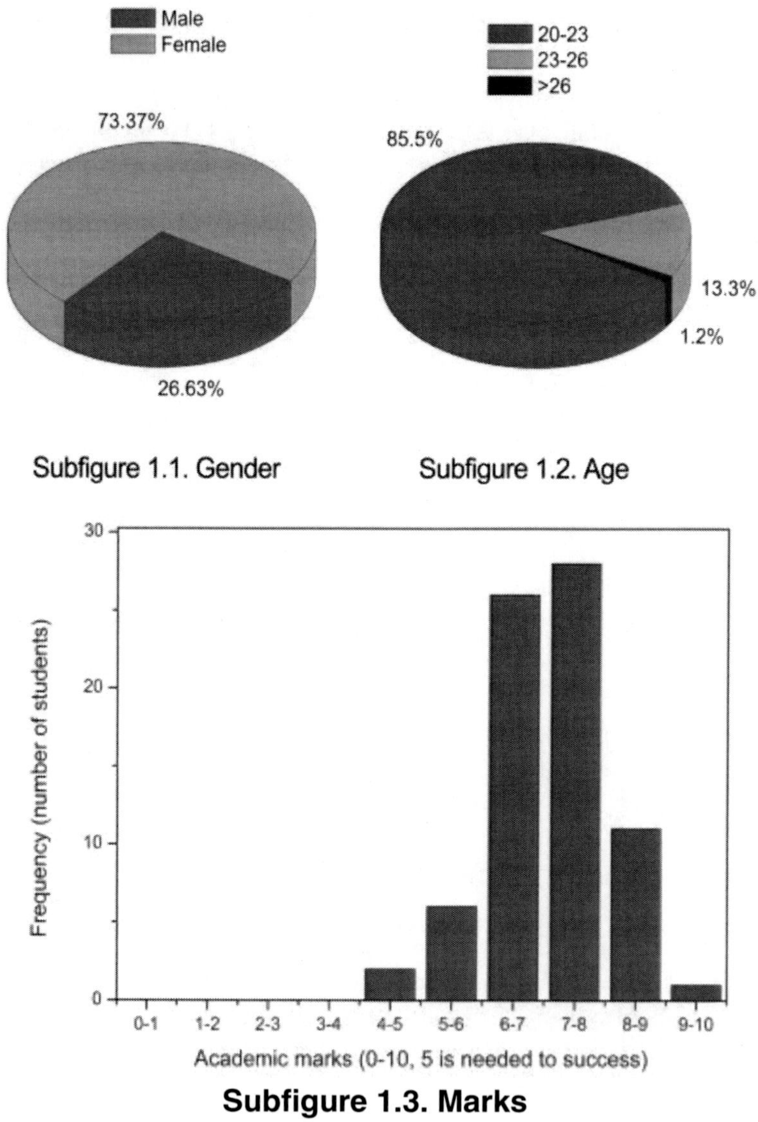

Figure 1. Demographic description of the sample.

The final questionnare was a two-pages document with 21 questions, included in four main groups: socio-demographic data, emotions towards the contents in the matter, emotions towards the teaching method and origin and causes of such emotions. Every question was graded according to a Likert-like scale (0-9). This questionnare was adequately validated with experts.

Students were asked to fill up this questionnarire choosing which emotion was significatively felt. The choices included happyness, confidence, joy, admiration, quietness, satisfaction, enthusiasm, suprise and amusement as positive emotions. Possible negative emotions were concern, embarrassment, anxiety, fear, disgust, sadness, wrath, nervousness and boredom. Each subject selected the most representative emotion for each content unit and the level they felt it. The same was done on the didactic methodology used by the teacher.

2.3. Some Results

Since each student can fill up the questionnaire with different emotions and different degrees of feeling, the results must be a data simplification. In this sense, Figure 2 presents the percentage of students that have filled up the main emotion with an average high level.

As can be seen, the narrative units (Astronomy and Energy sources) are linked to positive emotions while the rest of units (where calculus and mathematics are involved) presented a prevalence of Boredom feeling. In fact, when asked about the teaching methodology used for explaining each syllabus unit, students responded that the prevalent emotion is, again, boredom. No matter what the teacher does in the classroom, boredom is the universal feeling amongst the students.

When contrasted the level of this emotion with the academic marks of the students, no correlation is obtained under statistic considerations. That is, Chi-squared test linking marks and boredom level was unespecific in the large majority of the questions. Consequently, one can reflect the intrinsic nature of the matte is linked to a prevalent boredom emotion, possibly due to the scientific vocation of the pre-service teachers.

These results obviously urged us to implement new strategies for overcoming boredom, since this negative emotion is perhaps the most harmful for the scientific education.

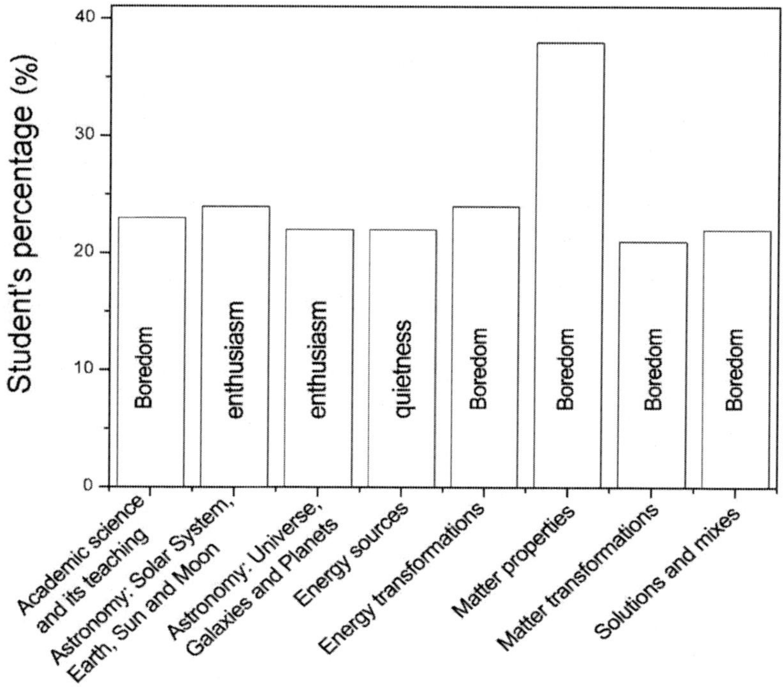

Figure 2. Main emotions linked to main syllabus units in Matter and Energy Education.

3. Innovation for a Better Science Learning Experience: The Case of Gardner's Multipe Intelligences Theory

Gardner's Multiple Intelligences Theory (MIT) arose more than 30 years ago (Gardner, 1983) and does not limit intelligence to its purely cognitive facet (Gardner, 1995), but takes account of its affective and emotional sides. Despite of it age, it is difficult to find specific educative applications of such theory. The Intelligence Style (IS) of each subject is obviously relevant for designing the different learning strategies, but perhaps a twist on this could be the identificaton of the emotional performance of each subject according to his/her IS. This information will be useful for complementing the teaching process by considering the emotional response and consequently giving the student an opportunity to improve his/her academic performance. This was also reported by Gardner (1989) in the early years of the Multiple Intelligences

Theory (MIT) and this fact has been confirmed by other researchers (Al-Balhan, 2006).

Since Gardner set out his theory, many studies have been focused on the estimation of the prevalent thinking style (Furnham et al., 2005), but there exists a general consensus inside the scientific community on the adoption of different mechanisms for measuring the intelligence of the individuals, mainly by considering this measure as a way of enhancing one's own development rather than imposing certain limits to either personal and academic growth (Almeida et al., 2010).

The use of MIT inside the academic field as a teaching path for enhancing student's capabilities have been previously reported in some published papers with general purposes (Armstrong, 2009; Akkuzu and Akçay, 2009) or particular ones, that is, focusing the science and mathematics education (Sulaiman et al., 2010; Karamikabir, 2012).

Some authors have pointed out the relevance of incluiding MIT in the general orientation of academic Science and Technology lessons (McKenzie, 2005). In the same researching line, the current work has the objective of identifying and measuring the eventual good effect of including MIT-based activities in the development of a specific topic inside the Technology lessons. Consequently, a Didactic Unit was implemented where
up to 9 activities were involved, each one oriented for each IS. The aim of this work is to evaluate the double performance of such teaching methodology: cognitive and emotional. In other words, our working question was:

> Do these strategies improve the emotional and cognitive yield of the students?

Additionally, since we have identified the IS of the students, we can see if this teaching strategy benefits even more to some IS.

3.1. Sample Description

The study universe involved 157 subjects, splitted in two groups (control and experimental one). 53% of them belonged to the experimental group whereas 47% was assigned to the control one. The study involved 69 (44%) boys and 89 (56%) girls, and up to 4 Secondary Schools were implied in the work. Control group was submitted to traditional teaching procedures (magistral lessons and exposition by the teacher), while the experimental group received a more innovative teaching strategy, including the development of a Didactic Unit that was designed with up to 9 different activities, one for each IS.

3.2. Methodology

Since the current research merges two aspects of the teaching-learning process (the emotional yield and the content acquisition), three elements have been taken into account for data collection:

a) The tool for measuring the prevalent IS was adapted from previous works (Armstrong, 2009) which were mainly based on Gardner's first approaches (Gardner, 1983; 1989). The version designed by Giorgis (Giorgis, 2007) was used since it shows a wide acceptation amongst the spanish audience (Barrientos-Jiménez et al., 2009). In it the student must score in a 0-5 scale of agreement a total of 32 questions about the way he or she learns in an easier way. The analysis of such items should lead us to assign him/her a predominant IS.
b) In the case of the emotional score, the questionnaire was made *ad hoc*. In it, students were asked to write down what emotions feel when doing a particular task. For example, when asked about the feelings when studying the effects of electricity, a student could answer 'surprise'. The answer choices, regarding the involved emotions, were Surprise, Happyness, Sadness, Wrath and Fear. Each emotion has a specific score from 10 (Surprise) to 2 (Fear). The global score gave us the positive emotional score for each student. A maximum of 100 points can be obtained. This scoring is built up according to some recent studies and statements, e.g., Gianotti (2015) declared to a national newspaper that *The maximum satisfaction for a scientist is surprise*. In the academic field, it is well-known the link between suprise and exploration aim (Plutchik, 2001), that was recently revisited by Choliz (2005). Experimentally, Conejo et al. (2015) have proved that technolgy-assisted gaming is positive for enhancing the knowledge acquisition in science and for promoting the scientific skills development through surprise and entertainment activities.

c) Finally, the contents acquisition test was the specific evaluation exam of the topic. The maximum score that can be obtained here is again 100.

Each test was made before and after the teaching intervention, and the contrast was done on an Experimental group and on a Control group, so the differences between them revealed the efficiency of the teaching methodology. The three questionnaries were adequately validated and their robusticity and internal coherency were checked out.

Data analysis was carried out by using SPSS package. Technically, the statistical analysis results were contrasted with an $\alpha = 0.5$ (confidence p-value of 0.05).

3.3. Some Results

As an initial approach to the data results, a preliminary test for confirming the validity of the whole study must be carried out on the data set. This consists of the confirmation of equal variances (Levene's statistics) and a one-way ANOVA statistical test for accepting the hypothesis of undistinguisibility. This means the variable 'Group' (Empirical or Control) does not affect to the preliminary value of the mean emotional or knowledge score and consequently both groups are undistinguisable in terms of previous emotinos or knowledge content.

The statistical results of this first step confirmed our working hypothesis. Levene's statistical reached 0.05 (p-value of 0.82) in the case of knowledge content and 1.1 (p-value of 0.29) in the case of emotional score. This confirms equal variances. Additionally, ANOVA test gave a p-value of 0.08 for the first analysis (knowledge) and 0.85 for the second one (emotional rate). Both of them are over 0.05, the significativity limit, so the preliminary hypothesis of undistinguisability can be statistically accepted. Both groups are consequently homogeneous for the intervention analysis.

The second confirmation we attempt to reach is the assumption that both educative interventions (the traditional one and the MIT-based one) do effect on the emotional and knowledge acquisition yield of the students. This can be observed through a t-Student statistical test, which will give us if there exists any statistically significant difference between the scores obtained by the pre-test and those obtained by the post-tests. The statistical significance of t-

Student test was given by a p-value of 0.00, so the differences due to the educative intervention (both kinds) are evident.

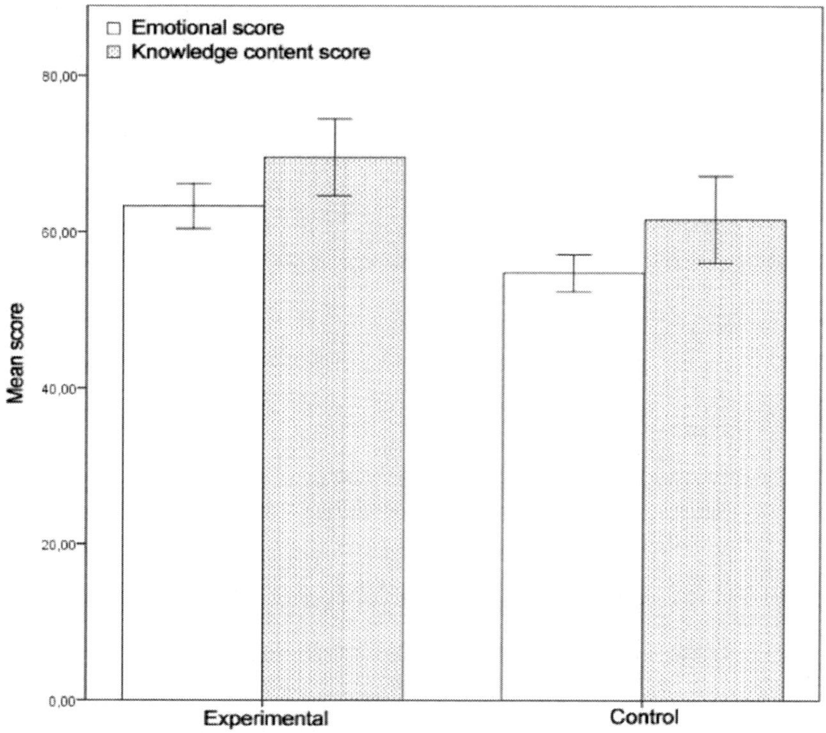

Figure 3. Mean score for emotional and knowledge content in post-test evaluation for Experimental and Control group. Confidence interval for error bars of 95%.

Although both teaching methodologies have been checked in terms of effect on the students' emotional and coginitive yield (being favourable), as the study universe was confirmed to be homogeneous, the post-test results can be compared in order to establish which teaching strategy leads to a better results, either in emotional and cognitive aspects. This can be easily carried out through a one-way ANOVA test where we can compare the results of both post-test in the control group and in the experimental one. ANOVA est also gives us the statistical data of a p-value of 0.00 for emotional performance and 0.34 for content acquisition (cognitive yield). Both p-values are under the significativity limit of 0.05, so one can affirm with a 95% of confidence that there exists differences between the two groups in terms of emotional and cognitive yield, being more favourable in the experimental group. In other words, the results obtained in the experimental group by using the MIT-based teaching strategy are better than those obtained in the control group with the traditional exposition methodology. The comparison can be seen graphically in Figure 3.

Figure 3 clearly depicts the emotional and cognitive performance. When compared blank columns and filled ones amongst them, error bars are clearly separated in the case of emotional score and slightly overlap in the case of content acquisition. The p-values confirmed the statistical significance of such results.

CONCLUSION

Students react in different ways when they are subjected to a teaching-learning process and these emotional responses are different enough to take them into account and try to enhance them. Scientific matters usually are not favoured in pre-service teachers, since they make boredom emotion to arise whatever the teacher does in classroom. This is the reason science is claiming for novel and innovative methodologies that empowers surprise, enthusiasm and happyness in the students.

This work has proven the benefits of including a specific attention to the Intelligence Styles when teaching circuitry, a particularly hard syllabus unit in secondary education. Either the content acquisition and the emotional performance of the students subjected a statistically significant raise with this methodology, where every student were suggested to complete a list of 9 activities, one for each Intelligence Style. The statistical analysis of this educative intruction revealed that, compared with a control group, the students that carried out these activities reached higher emotional yield and their cognitive performance was also higher than those who followed the traditional transmission method.

These results should encourage teachers for paying special attention to Gardner's Multiple Intelligence Theory when scheduling their teaching activities, since a particular focus on each Intelligence Style enhances not only the emotional performance of the whole group, but also the content acquisition in the involved students.

ACKNOWLEDGMENTS

This work was partially supported by the Spanish Government (Economy and Competitivity Ministry) under EDU2012-34140 project, as well as by the Regional Government of Extremadura and by the Researching Group DEPROFE, from University of Extremadura.

REFERENCES

Akkuzu, N. and Akçay, H. (2011) The design of a learning environment based on the theory of multiple intelligence and the study its effectiveness on the achievements, attitudes and retention of students Procedia *Computer Science*, 3, 1003-1008.

Al-Balhan, E. (2006) Multiple Intellingences Styles in Relation to improved academic performance in Kuwaiti Middle School reading *Digest of Middle East Studies*, 15, 18-34.

Almeida, L.; Prieto, M.; Ferreira, A.; Bermejo, M.; M. Ferrando and Ferrándiz, C. (2010) Intelligence assessment: Gardner multiple intelligence theory as an alternative *Learning and Individual Differences*, 20, 225-230.

Álvarez, G. and Canal, J. Mellado, V.; Blanco, L.; Borrachero, A. and Cárdenas, J. (Eds.) (2013) *Perception of the student's emotions on Technology lessons in Secondary Obligatory Education* (*Percepción de las emociones en el alumnado de tecnología de Educación Secundaria Obligatoria*) Emotions in the Science and Mathematics teaching and learning process (Las emociones en la enseñanza y el aprendizaje de las ciencias y las matemáticas), DEPROFE, 484-493.

Armstrong, T. (2009) *Multiple Intelligences in the Classroom*, 3rd Edition Association of Supervision and Curriculum Development, London.

Arribas, M. (2004). Design and Validation of Questionnaries (Diseño yvalidación de cuestionarios) *Matronas profesión*, 5 (17), 23-29.

Barrientos Jiménez, E.; Mattza Díaz, I.; Vildoso Villegas, J. and Carola Sánchez, T. (2009) Multiple intellingences, learning styles and accomplishment level (Las Inteligencias múltiples, los estilos de aprendizaje y el nivel de rendimiento) *Investigación Educativa*, 13, 9-19.

Beauchamp, B. and Parkinson, J. (2008) Pupils' attitudes towards school science as they transfer from an ICT-rich primary school to a secondary school with fewer ICT resources: Does ICT matter? *Education and information technologies*, 13, 103-118.

Bisquerra, R. (2001) *Emotional education and welfare* (*Educación emocional y bienestar*) Wolters Kluwer, The Netherlands.

Borrachero, A.; Brígido, M.; Gómez, R.; Bermejo, M. and Mellado, V. (2011) Emotions in prospective secondary teachers on the science learning and teaching process (Las emociones de los futuros profesores de secundaria sobre el aprendizaje y la enseñanza de las ciencias) *International Journal of Developmental and Educational Psicology*, 23, 521-530.

Black, P. and Atkin, M. Taylor and Francis (Eds.) (2005) *Changing the subject*: *Innovation in Science, Mathematics and Technology Education.* Organization for Economic Cooperation and Development.

Brígido, M., Caballero, A., Bermejo, M. L., Conde, C. and Mellado, V. (2009). Emotions in Science students in preservice teachers (Las emociones en ciencias de estudiantes de Maestro de Educación Primaria en prácticas) *Campo Abierto*, 28(2), 153-177.

Brígido, M.; Borrachero, A.; Bermejo, M. and Mellado, V. (2013a) Prospective primary teachers' self efficiacy and emotions in science teaching *European journal of teacher education*, 36, 200-21.

Brígido, M.; Couso, D.; Gutiérrez, C. and Mellado, V. (2013b) The emotions about teaching and learning science: a study of prospective primary teachers in three spanish universites *Journal of Baltic Science Education*, 12, 299-311.

Casacuberta, D. *What is an emotion*? (2000) Critica, Barcelona, Spain.

Choliz, M. (2005) *Emotion psychology. The emotional process.* On-line autopublication.

Conejo, R.; Garcia-Vinas, J.; Gaston, A. and Barros, B. Technology-Enhanced Formative Assessment of Plant Identification *Journal of Science Education and Technology*, Springer Netherlands, 2015, 1-19.

Costillo Borrego, E., Borrachero Cortés, A. B., Brígido Mero, M. and Mellado Jiménez, V. (2013). Emotions on science and mathematics teaching learning in Secondary preservice teachers (Las emociones sobre la enseñanza-aprendizaje de las ciencias y las matemáticas de futuros profesores de Secundaria) *Revista Eureka sobre Enseñanza y Divulgación de las Ciencias*, 10, 514-522.

Damasio, A. (2010) *And the brain created man* (Y el cerebro creó al hombre) Destino, Madrid.

Dos Santos, F. and Mortimer, E. (2003) How emotions shape the relationship between a chemistry teacher and her high school students *International Journal of Science Education*, 25, 1095-1110.

Duit, R.; Treagust, D. and Widodo, A. Vosniadou, S. (Ed.) (2008) *Teaching science for conceptual change: theory and practice* International handbook of research on conceptual change, Routledge, 629-646.

Fernández, M. J., Brígido, M. y Borrachero, A. B. (2013). *Longitudinal study on emotions and actitudes in preservice teachers towards the experimental science teaching (Estudio longitudinal sobre las emociones y actitudes del alumnado de maestro del Grado de Educación Primaria ante la enseñanza de ciencias experimentales*). In: Mellado, V.; Blanco, L.; Borrachero, A. and Cárdenas, J. (Eds.) Emotions in the Science and Mathematics teaching and learning process (Las emociones en la enseñanza y el aprendizaje de las ciencias y las matemáticas), DEPROFE, 2, 351-371.

Fernández-Abascal, E.; Martín, M. and Domínguez, J. (2001) *Psichological process* (Procesos psicológicos) Pirámide, Spain.

Francisco, V.; Gervás, P. and Hervás, R. (2005) Analysis and synthesis of the emotional expression in read-aloud tales (Análisis y síntesis de expresión emocional en cuentos leídos en voz alta) *Natural language process* (*Procesamiento del lenguaje natural*) 35, 293-300.

Furnham, A.; Wytykowska, A. and Petrides, K. (2005) Estimates of Multiple Intelligences: A Study in Poland *European Psichologist*, 10, 51-59.

Gardner, H. (1983) *Frames of Mind*: *The Theory of Multiple Intelligences* Basic Books, London.

Gardner, H. (1995) *Reflections on multiple Intelligences: Myths and Messages* Phi Delta Capan, Phi Delta Capan, 77, 200-209.

Gardner, H. (1989) Educational implications of the Theory of Multiple Intelligences *Educational Researcher*, 18, 4-10.

Gianotti, F. (2015) *There is no difference between arts and science*, EL PAIS, Spain.

Giorgis, N. (2007) Multiple Intelligences Profiles (Perfil de Inteligencias Múltiples) *Electronic Bulletin 'Engineering First' (Boletín Electrónico 'Ingeniería Primero'*).

Goleman, D. (1996) *Emotional intelligence* Bantam Books, London.

Hargreaves, A. Hargreaves, A.; Lieberman, M.; Fullan, M. and Hopkins, D. (Eds.) (1998) *The emotions of teaching and educational change* International handbook of educational change, Kluwer Academic Publishers, 558-575.

Hill, A. and Smith, H. (1998) Practice meets theory in Technology Education: A case of authentic learning in the High School Setting *Journal of Technology Education*, 9, 29-45.

Karamikabir, N. (2012) Gardner's Multiple Intelligence and Mathematics Education Procedia, *Social and Behavioral Sciences*, 31, 778-781.

Lavonen, J.; Gedrovics, J.; Byman, R.; Meisalo, V.; Juuti, K. and Uitto, A. (2008) Students' motivational orientations and career choice in science and technology: a comparative investigation in Finlad and Latvia *Journal of Baltic Science Education*, 7, 86-102.

Manassero, M. (2013) *Emotions: from omission to relevancy in the behavior explanation* (*Emociones: del olvido a la centralidad en la explicación del comportamiento*) In: Mellado, V.; Blanco, L.; Borrachero, A. and Cárdenas, J. (Eds.) Emotions in the Science and Mathematics teaching and learning process (Las emociones en la enseñanza y el aprendizaje de las ciencias y las matemáticas), DEPROFE, 2, 3-18.

Marbá-Tallada, A. and Márquez, C. (2010) What do students think on Science lessons? A transversal study from Sixth Degree (Primary School) to Fourth Degree in Secondary Obligatory Education (¿Qué opinan los estudiantes de las clases de ciencias? Un estudio transversal de sexto de primaria a cuarto de ESO) *Enseñanza de las ciencias*, 28, 19-30.

McKenczie, W. (2005) *Multiple Intelligences and instructional technology* International Society of Technology Education, ISTE Publications, New York.

Otero, M. (2006) Emotions, feelings and reasoning in science education (Emociones, sentimientos y razonamientos en didáctica de las ciencias) *Electronic Journal of Science Education* (*Revista electrónica en educación en ciencias*), 1.

Pérez, A. and De Pro (2013) *A demoscopy study on the feelings and thinks children and adolescent about the formal science education (Estudio demoscópico de lo que sienten y piensan los niños y adolescentes sobre la).*

Pintrich, P. R.; Marx, R. W. and Boyle, R. A. (1993). Beyond cold conceptual change: the role of motivational beliefs and classroom contextual factors in the process of conceptual change. *Review of Educational Research*, 63(2), 167-199.

Plutchik (2001) The nature of emotions *American Scientist*, 89, 344-350.

Shapiro, S. (2010) Revisiting the teachers' lounge: Reflections on emotional experience and teacher identity *Teaching and Teacher Education*, 26, 616-621.

Sulaiman, T.; Abdurahman, A. and Rahim, S. (2010) Teaching Strategies Based on Multiple Intelligences Theory among Science and Mathematics Secondary School Teachers Procedia, *Social and Behavioral Sciences*, 8, 512-518.

Thagard, P. (2009) *Why Cognitive Science Needs Philosophy and Vice Versa* Topics in Cognitive Science, Blackwell Publishing Lt, 1, 237-254.

Van der Hoeven Kraft, K.; Srogi, L.; Husman, J.; Semken, S. and Fuhrmman, M. (2011) Engaging students to learn through the affective domain: A new framework for teaching in the geosciences. *Journal of Geoscience Education*, 59, 71-84.

Vázquez, A. and Manassero, M. (2007) In defense of attitudes and emotions in the scientific education (I): Evidences and general arguments (En defensa de las actitudes y emociones en la educación científica (I): Evidencias y argumentos generales) *Eureka Journal of Science Education and Divulgation* (*Revista Eureka sobre Enseñanza y Divulgación de las Ciencias*), 4, 247-271.

Vázquez, A. and Manassero, M. (2011) The decrease of children's attitudes towards the science in the obligatory education (El descenso de las actitudes hacia la ciencia de chicos y chicas en la educación obligatoria) *Ciência e Educaçâo*, 17, 249-268.

Weiner, B. (1986) *An attributional theory of motivation and emotions* Springer, The Netherlands.

Wu, H. and Huang, Y. (2007) Ninth-grade student engagement in teacher-centered and student-centered technology-enhanced learning environment *Science Education*, 91, 727-749.

In: Teaching and Learning
Editor: Malcolm Vargas
ISBN: 978-1-63485-228-9

Chapter 4

UNDERGRADUATE STUDENTS' HEURISTICS AND STRATEGY PATTERNS IN RESPONSE TO WEB-BASED PEER AND TEACHER ASSESSMENT FOR SCIENCE LEARNING

Olia E. Tsivitanidou [*] ***and Costas P. Constantinou*** [†]
Learning in Science Group, Department of Educational Sciences, University of Cyprus, Nicosia, Cyprus

ABSTRACT

Peer-assessment is an educational arrangement in which students are actively involved in assessing their peers' work; it can be perceived also as a learning activity due to the benefits which arise from its enactment. In this study, we sought to examine the heuristics and strategies of undergraduate student teachers in the roles of peer-assessor and peer-assessee in a web-based peer-assessment context as part of a science course. We also aimed to examine whether the strategies that we were able to identify associate with students' learning gains and also with the type of feedback produced and received, respectively, in the two roles. Participants were 24 undergraduate student teachers in Primary Education from a university in Cyprus, who were studying a web-based unit on marine Ecosystems. Students worked in pairs, but on different computers,

[*] Corresponding Author address, Email: tsivitanidou.olia@ucy.ac.cy.
[†] c.p.constantinou@ucy.ac.cy.

while enacting one inquiry cycle. Each step of this cycle led to the development of a learning artefact, which was uploaded by every student individually in their own web-portfolio; in total each student developed eight learning artefacts. By the end of the teaching unit, students reciprocally and individually assessed the web-portfolio of a peer-group, with the assistance of 3-Likert scale rubrics with pre-defined criteria. Students' web-portfolios were also assessed by the teacher using the same rubrics. The students were then allowed to revise their web-portfolios based on both the peer- and teacher- feedback received. Their actions on their screens were recorded during the whole intervention with the use of the River Past Software Pro. Data were collected from five sources; namely screen-video captured data, peer and teacher-feedback, pre- and post- instructional tests, pre- and post- instructional interviews and students' initial and revised learning products. Data were exposed to qualitative analysis and then also treated quantitatively through the use of non-parametric tests. Findings revealed two main assessor and three assessee profiles. In particular, assessors were distinguished into *autonomous* (profile 1) and *informed* (profile 2) and assessees into *teacher trusting* (profile 1), *teacher trusting and skeptical towards peer feedback* (profile 2) and *teacher and peer trusting* (profile 3). Overall, students' learning gains were not associated with their actions on either the peer-assessor or peer-assessee role; all students had positive learning gains at the end of the intervention, regardless of the strategies they had followed. However, assessors' profiles had statistically significant differences in the explicit guidance offered as part of the feedback produced; informed assessors offered more guidance compared to the guidance offered by autonomous assessors. In addition, assessees' strategies were found to be associated with the structural components of peer and teacher-feedback received. Time pressure was a critical factor that affected assessees' actions. The findings have implications for teaching practice as well as policy for teaching and learning in web-based learning environments.

Keywords: scaffolding in reciprocal peer assessment, heuristic patterns, peer-assessor profiles, peer-assessee profiles, web-based science learning

INTRODUCTION

Formative assessment is the type of assessment which occurs during the learning process and could be used to promote but also evaluate learning (Bell and Cowie, 2001; Pellegrino, Chudowsky and Glaser, 2001). A novel approach in formative assessment is the active involvement of students in assessing a peer's work, known as peer assessment (Harrison and Harlen, 2006; Harlen, 2007; Cestone, Levine and Lane, 2008). Peer assessment is defined as an educational arrangement where students judge peers' performance by providing grades, and/or offering written or oral feedback (Topping, 1998); it occurs among individuals who have experienced a common learning context (Topping, 1998) and subsequently have similar backgrounds (Falchikov and Goldfinch, 2000). It is worth considering as an assessment method, since it endorses many benefits (Boud, 1995; Brown, Bull and Pendlebur, 1997; Dochy, Segers and Sluijsmans, 1999; Topping, 2003) but also as a learning tool (Orsmond, Merry and Reiling, 1996). Previous studies have shown that peer-assessment could enhance students' reflection and in-depth thinking (Bouzidi and Jaillet, 2009; Chang, Tseng, Chou and Chen, 2011), since it requires students to use their knowledge and assessment skills to review, clarify, and correct other peers' work (Ballantyne, Hughes and Mylonas, 2002). Therefore, by enacting peer assessment, students are given the opportunity to become more critical, independent learners as they gain familiarity with the application of assessment criteria and the assessment process (Searby and Ewers, 1997; Falchikov, 1995) but also as they engage in reflection processes while assessing (Hwang, Hung and Chen, 2014) and reviewing the peer feedback received (Yang, 2011). Moreover, learners develop responsibility and a sense of accountability for the learning of their peers (Topping, Smith, Swanson and Elliot, 2000; Dochy, et al., 1999; Orsmond, et al.,1996) and, most important, their own learning (Davies, 2006; Weaver and Cotrell, 1986). As a result, students benefit from the peer-assessment procedure in two ways: while undertaking the role of the assessor and the assessee.

This type of reciprocal peer-assessment enables students to gain experience both in the task of assessing peers and in being assessed by peers. As peer assessors, students develop better insight on their own work, since the provision of feedback requires reflection processes (Prins, Sluijsmans, Kirschner and Strijbos, 2005; Xiao and Lucking, 2008); moreover, they have opportunity to improve their problem-solving skills, as well as their learning achievements (Cheng and Ku, 2009; Chen Wie, Wu and Uden, 2009; Duran and Monereo, 2005; Hwang, et al., 2014). As peer assessees, students can use the peer feedback received to improve their learning artefacts (Cheng and Ku, 2009; Duran and Monereo, 2005) and again develop their critical thinking skills while filtering peer feedback (Hovardas, Tsivitanidou and Zacharia, 2014). Overall, the peer-assessment method, as one example of formative assessment approaches, seems to improve students' learning achievements

(Falchikov, 2003; Harrison and Harlen, 2006) and performance (Black and Harrison, 2001; Sluijsmans, Brand-Gruwel and Van Merriënboer, 2002) in various domains and in particular in science education (Cho and Schunn, 2007; Chen, et al., 2009; Crane and Winterbottom, 2008; Tsai and Liang, 2009; Tsai, Lin and Yuan, 2002; Tseng and Tsai, 2007; Pare and Joordens, 2008; Ploegh, Tillema and Segers, 2009; Prins, et al., 2005; Tsivitanidou and Constantinou, 2016; Tsivitanidou, Zacharias and Hovardas, 2011).

Despite the potential benefits that peer-assessment might entail in enhancing students' congitive, meta-congitive and social affective development (Kollar and Fischer, 2010), peer-assessment remains a demanding task especially for novices and particularly when no scaffolds are offered. In order to facilitate peer-assessment, educators and researchers stress the importance of offering appropriate scaffolds that can enhance students' assessment skills, especially for those who are novices in peer-assessment (Hovardas, et al., 2014). A commonly used type of scaffolding is the provision of assessment criteria to peer-assessors, usually in the form of rubrics (see, for instance, Hafner and Hafner, 2003). The provision of assessment criteria increases the possibility of establishing a common understanding between students and the teacher of what has to be assessed (Orsmond, Merry and Reiling, 2000). However, prior research has indicated that even the use of assessment criteria may not necessarily guarantee sound psychometric measures (e.g., Tsivitanidou et al., 2011). An alternative approach in this direction, counter to the general trend in the literature, does not require assessment criteria; this approach is called the comparative judgment approach (CJA) (Jones and Alcock, 2014). The rationale for CJA is based on a long-standing psychophysical principle, usually called *"The Law of Comparative Judgement"* proposed by Thurstone (Thurstone, 1927, as cited in Jones and Alcock, 2014). In this context, peer-assessors are given pairs of student artefacts and are requested to judge which artefact is "better"; and therefore, which one of the two assessees has better performance (Pollitt, 2012). In fact, it has been reported that students may use this approach implicitly by themselves, even when they are not instructed to do so (Tsivitanidou and Constantinou, 2016).

For determining the type and/or degree of scaffolds that students need overall and in particular while peer-assessing, it is essential to obtain insight into the needs that arise when students get involved in such tasks. Hwang et al. (2014) have proposed as a topic worthy of further investigation, the examination and analysis of heuristics and actions of students when enacting peer assessment. Few studies have developed a focus on examining students' actions and strategies, when engaged in peer-assessment (Hovardas et al., 2014; Tsivitanidou and Constantinou, 2016).

We have only been able to identify these two studies in the literature that have examined students' actions and strategies while enacting computer-supported peer-assessment in science education. The first study of Hovardas et al. (2014) examined secondary school students' behavioral profiles in supported and structured peer-assessment. The second study of Tsivitanidou and Constantinou (2016) has explored students' profiles in unsupported and unstructured reciprocal peer-assessment, as strategy patterns that peer-assessors and peer-assessees follow respectively. In the latter study, the relation of those profiles with students' learning gains was also examined. In both studies, it has been shown that peer-assessors might return to their own learning products, while providing peer feedback, which can be used as a point of reference while assessing artefacts developed by their peers. As already stated, this strategy resembles the CJA, initially proposed by Thurstone (Thurstone, 1927, as cited in Jones and Alcock, 2014). Considering that students in the Hovardas et al. study were also provided with scaffolds while peer-assessing (3-Likert scale rubrics with a wide range of assessment criteria), it is quite interesting to see that, even in that case, students might need to search for additional sources of information and points of reference, for accomplishing the assessment task.

With respect to the assessee role, the two aforementioned studies differentiate from another in two terms: (i) the type of peer-feedback received in each case (feedback in the form of rubrics in the Hovardas et al. study, written comments in the Tsivitanidou and Constantinou study); (ii) the person from whom assessees received feedback (teacher and peer feedback in the first study, only peer-feedback in the second study). In particular, in Hovardas et al. (2014) students were distinguished into four profiles: (i) peer assessees who studied both teacher and peer feedback, but used the feedback that they had produced as peer assessors to filter it, before making any change (profile 1); (ii) peer assessees who read both teacher and peer feedback and considered the teacher feedback to be more valuable, but in the end did not proceed with any changes in their learning artefacts (profile 2); (iii) peer assessees who proceeded with revisions in their learning artefacts, taking into consideration both teacher and peer feedback (profile 3); and (iv) peer assessees who quickly scanned both teacher and peer feedback and then concentrated only on teacher feedback, which they partially filtered through the use of the feedback they produced for others as peer assessors before proceeding with revisions to their learning artefacts (profile 4). The second study of Tsivitanidou and

Constantinou (2016), revealed three distinct patterns of peer assessees' typical responses and actions while receiving peer feedback: *engaged but unlucky* (profile 1), *disinterested* (profile 2) and *engaged* assessees (profile 3). The three profiles differed from each other regarding the following actions: viewing one's own learning product, for which one requested feedback and studying the feedback received (present in the 2nd category of the 2nd profile and the 3rd profile), viewing one's own learning product, for which one requested feedback and checking for possible feedback received (present in the 2nd category of the 1st profile only), editing one's own learning product after its initial creation (present in the 1st category of the 1st profile and in the 3rd profile). Overall, students who were fortunate to have received peer-feedback and who had also positive attitudes towards peer-feedback, thus devoting time to review it, where those who had higher learning gains.

In this study we wanted to further explore how peer-assessor typically act while offering peer-feedback via the use of rubrics, how peer-assessees typically act when receiving peer- and teacher feedback and how those actions might associate with students learning gains in the teaching unit of ecosystems. With this study, we sought to take research in this field one step further, in an effort to understand how students typically act and what strategies they use in reciprocal peer-assessment.

Objectives and Research Questions

The purpose of this study was to examine peer-assessor and peer-assessee profiles, patterns of student strategies and actions that emerge when undergraduate student teachers undertake the peer assessor and peer assessee role respectively, in reciprocal web-based peer-assessment. Also, we aimed to investigate how those profiles might relate with students' learning gains in the topic under study. Due to the fact that peer assessment can be operationalized in many forms (e.g., paper-based or computer-based, offline or online) and types (e.g., one-way or reciprocal, supported or unsupported, structured or unstructured), any research in the peer assessment domain should be well defined, both in terms of its form and type, in order to avoid confounded conclusions (Topping, 1998; Van Gennip, Segers and Tillema, 2010). In this study, we employed a reciprocal (for examining both roles of assessor and assessee), anonymous (for ensuring highest possible level of privacy and, hence, providing students with the opportunity to provide, in a more objective manner, critical reviews to their peers), and online (asynchronous) peer assessment context (for supporting the mechanisms of reciprocal peer assessment via provision of isolated channels of communication between pairs, anonymity, provision of a wider range of learning objects to be produced and to be assessed) (Tseng and Tsai, 2007; Xiao and Lucking, 2008). From a methodological perspective, the employment of peer-assessment via the use of computers, facilitated also the collection of a rich pool of data, providing us with the opportunity to record and keep track of students' actions as displayed on the monitor. Furthermore, *Reciprocal Peer-Assessment* as employed in this study was *Supported* and *Structured* (SSRPA). With the term *supported* we refer to the peer-assessment method implemented with the provision of at least one scaffold that could enhance the assessment task (e.g., assessment criteria etc.), while as *structured* we define the peer assessment method in which the pairs of assessors-assessees are predefined by the teacher, and students enact peer-assessment at a specific phase of the teaching intervention again determined by the teacher. The curriculum material used in this study was inquiry-based, focused on ecosystems and was available through a web platform.

Our research questions were:

1) What are the various peer-assessor and peer-assessee action patterns and profiles in SSRPA?
2) Are the emerging peer-assessor / assessee profiles associated with students' learning gains in the chosen teaching unit on ecosystems?

METHODS

Participants

Participants were 24 undergraduate students of primary education from a university in Cyprus. They were all fourth-year student teachers participating in a science specialization course. The low sample size helped us collect and analyse a large amount and a broad variety of data to triangulate the answers to the research questions. The participants were guaranteed anonymity and that the involvement in peer-assessment activities would not contribute to their final grade at the end of the semester. None of these students had prior experience with SSRPA as implemented in this study. The students worked in pairs throughout most parts of the intervention: students were matched randomly and the group compositions did not change during the intervention.

However, students worked individually and anonymously with the assistance of scaffolds in enacting the peer assessor role.

Material

Students worked though web-based material that was developed for the purposes of the Science Created by You (SCY) project (http://www.scy-net.eu/) (de Jong et al., 2010; 2012). The topic was 'marine ecosystems' (Pedaste, de Jong, Sarapuu, Piksööt, van Joolingen and Giemza, 2013) and was aligned with the syllabus of the specific university course for science specialization of future primary teachers. The SCY-Lab learning environment follows a pedagogical approach that emphasizes learning artefacts (emerging learning objects), which are created by the students themselves. In this study, the students studied one of the socio-scientific 'missions' offered in SCY-Lab (the ECO mission) that was designed to guide them in a process of identifying general principles underpinning the functions of ecosystems. Students can visualise in the SCY-Lab learning envoriment each mission via the so called "mission map". The ECO mission is organized into four topics that can be enacted separately. For the purposes of this research, students studied part of this mission and were asked to complete one of the four designed inquiry cycles, which focuses on a problem involving prey-predator relationships.

At the beginning of the intervetion the participants were asked to construct a concept map, using a tool embedded in the SCY-Lab learning environment, brainstroming their initial ideas around fresh-water ecosystems. Then they initiated the ECO mission. A story was given about two children who went fishing in a lake and were confronted with a paradox situation: they observed that pikes (predator species) might outnumber roaches (prey species). Participants were asked to provide an explanation for this observation. They were allowed to do some background reading, from a list of texts that were embedded in the SCY-lab environment and a series of websites available, concerning relationships between trophic levels in an ecosystem, characteristics of aquatic ecosystems, and prey-predator interactions. While doing the background reading, the students were prompted to focus on the parameters involved in prey-predator interactions (namely, population sizes of the prey and predator populations, and trends of prey and predator populations in time). After becoming familiar with the most important aspects of these phenomena, the students were asked to formulate the main problem to be addressed. Students had to consider that the formulation of the problem (i.e., children's' unexpected observation) should not involve any type of human intervention as a cause, since human interventions had been excluded as a cause of the observed phenomenon during the discussions the two children had with local fishermen. The next step of their investigation involved experimentation with a modelling tool (SCYDynamics). Values characterizing predator and prey in an ecosystem could be used as either input or output variables in the SCYDynamics tool. Specifically, population gains and losses, ability to catch and feed on prey, natural growth and death factors, and the ability to convert prey biomass to predator biomass are specific elements in this simulation. For example, students could change the initial level of biomass of the components and then investigate changes for a specified time period (Pedaste, et al., 2013). Beforehand, students were asked to formulate two research questions and two corresponding hypotheses. In addition, students were asked to describe the various actions they would like to perform in order to run their experiment in the modeling tool provided; the resulted learning artefact was the experimental procedure. After having accomplished all the above-mentioned activities, students started working with the modeling tool provided. As the next step, students were asked to interpret their findings and formulate a solution to the problem they had defined at the beginning of the procedure.

The activity sequence was grounded in collaborative inquiry learning, including structured student collaboration in several steps. Overall, it took the student groups eight meetings of 1.5 hours each to complete this activity sequence. Instruction was enacted by two science teachers with prior peer-assessment and extensive science teaching experience, who attended preparatory meetings before each classroom session. The web-based peer assessment process was implemented via a networked file sharing system, hosted on the Dropbox™ platform (www.dropbox.com), which offered cloud storage and file synchronization. This web-based platform was chosen because the participants were already familiar with it and because it had adequate features for enacting the web-portfolios that were necessary for the purposes of the study.

Procedure

Before the intervention, the students signed a declaration of consent for their screen actions being recorded and for the interviews. The students were interviewed about their attitudes towards peer assessment. At the beginning of the intervention, all students completed a pre-instructional test on prior knowledge of concepts about prey-predator relationships. Then, the students started working on the ECO mission (prey-predator relationships) in the SCY-Lab platform, in groups of two, with each student working on a different computer and having her/his own web-portfolio for storing the learning artefcats s/he was creating (see Figure 1). Each task corresponded to the development of a learning artefact in the students' web-portfolio. Once the students had completed the ECO mission and created all the learning artefacts in the SCY-Lab platform, they were then asked to upload them in the Dropbox platform in a web-portfolio that was created for each student separately. All web-portfolios should have included eight learning artefacts at the end of the mission: (1) concept map about ecosystems; (2) problem formulation; (3) research questions; (4) hypotheses; (5) research design (6) data from SCYDynamics (7) inferences and (8) problem solution. Students' web-portfolios were coded to ensure anonymity.

As soon as students had completed their tasks and uploaded their learning artefacts into their own web-portfolios, they could access the web-portfolio of the peer group they were to assess, which was randomly assigned to them. Even though students worked in groups of two while studying the teaching material, they carried out the role of peer assessor on an individual basis, which allowed the production of two separate sets of peer feedback. More feedback carries a number of benefits, for example, provision of feedback on aspects of a learning artefact that the other assessor failed to address. Each web-portfolio (all the learner artefacts included in a web-portfolio) was assessed by two peers from the same group, who worked on different computers. Each web-portfolio was also assessed by one of the teachers, involved in this study. All of the web-portfolios were also assessed by the second teacher with similar qualifications, for internal reliability purposes (that implies that students received teacher from only one teacher). Cohen's kappa was calculated for all assessment criteria and all coefficients were over 0.88, which indicates that the two teachers produced feedback that was strongly consistent.

Peer assessors and teachers used a 3-Likert scale rubric with pre-specified assessment criteria. The assessment rubric was constructed by a panel of expets on the specific learning module completed, including two experienced science educators, an environmental scientist and an expert on scoring rubrics and measurement. The expert panel first went through learning objectives, curriculum material and the procedure and then identified specific attributes that should have been present in students' learning artefacts. This procedure resulted in a list of 43 assessment criteria grouped into two main categories: (1) criteria addressing each learning artefactindividually (n = 39), (2) criteria addressing the web-portfolio as a whole (n = 4) (all assessment criteria are presented in the Appendix). Assessors rated student performance on all criteria according to a 3-point Likert scale (i.e., (1) unsatisfactory; (2) moderately satisfactory; (3) (fully) satisfactory). Along with ratings, assessors were instructed to provide written feedback (for each criterion separately) to assessee groups, in which they were to explain the reasoning behind their ratings, provide judgments and suggestions for revisions. On average, it took each peer assessor about an hour to complete the assessment and for the teacher about twelve hours to assess all twelve web-portfolios from the study.

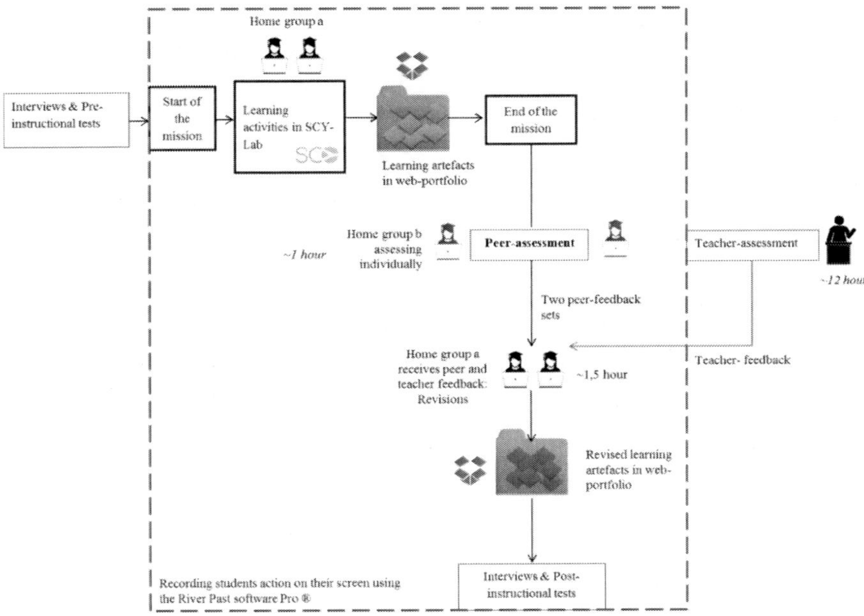

Figure 1. The reciprocal web-based peer-assessment (SSRPA) procedure. Rhombuses represent learning artefacts; folders stand for web-portfolios. Students start the mission and collaboratively do all the learning activities in pairs but on different computers; thus each student creates hers/his own learning artefacts in the SCY-Lab learning environment, and then uploads them in hers/his own web-portfolio in the Dropbox platform. During the peer review and assessment phase students work alone, whereas, during the revision phase they work in the same pairs as before in which they receive two peer-feedback sets and on teacher-feedback set. The students of home group a and b undertake both the role of the assessors and that of the assessee, even though it does not show in this figure.

As soon as the peer and teacher feedback were ready and sent to the corresponding assessee group, all participants switched to the role of the assessee. The revision phase in the assesse role took place in the subsequent course, so as to achieve the completion of the teacher assessments that lasted longer. Students in the peer assessee role, returned back to their initial home groups, processed the peer and teacher feedback received and decided on the actions to be taken. All assessments were sent from assessors to assessees automatically, through the networked sharing folder system embedded in the Dropbox platform, without revealing the identity of the peer assessor. However, assessees were aware about which feedback set corresponded to the peer-feedback and which to the teacher feedback. By the end of the revision phase, the students completed a post-instructional test, on their knowledge of concepts about prey-predator relationships, and were interviewed again about their experiences and attitudes towards reciprocal peer assessment. During the whole intervention, students' actions on their computer screens were recorded using screen capture software (River Past Screen Recorder Pro).

Data Collection

The data collection process involved five data sources: (1) screen-video-captured data, (2) peer and teacher feedback (3) pre- and post-instructional tests, (4) pre- and post-instructional interviews with students, (5) and students' initial and revised learning products. The peer and teacher feedback was stored on the server that hosted the Dropbox platform, along with the corresponding log files which included information necessary for our analysis, such as the code of the peer assessor, the date and time the feedback was created, revised, revisited, etc. Screen captured data were collected throughout the study with computer screen capture software (River Past Screen Recorder Pro). Screen recording allowed the collection of a rich record of actual computer work activity (e.g., actions and movements that took place on the computer monitor) in its natural work setting, showing the users' movement among various parts of the web-based study's material. With the assistance of a microphone and a camera, the software also allowed videotaping of the students in conjunction with what was taking place on the screen. Screen captured data were collected for all class meetings. For this study we have isolated particular episodes of the screen captured data. The isolated episodes concerned milestones were students were enacted peer-assessment both while giving feedback but also while receiving peer and teacher feedback and proceeding to revisions of their web-portfolio. All these episodes were coded for the on-screen actions/responses of peer assessors and peer assessees separately. The idea was to capture students' actions/responses while they enacted the roles of peer assessor and assessee.

Moreover, we used the pre- and post-instructional tests which were developed by the teachers in collaboration with a team of researchers in environmental education. The pre- and post- tests were identical and consisted by six open-ended questions which aimed to evaluate students' knowledge about the ecological pyramid and the dynamic relation between prey and predators. The tests were verified for their face and content validity by the work-team.

In regards to the interviews, a semi- structured protocol that consisted of 16 open-ended questions was used. The first question of the pre-instructional interviews was enquiring students whether they know "*what peer-assessment is.*" Then students were questioned about any previous experience in peer-assessment and its potential benefits. Moreover, students were asked about how capable they consider themselves for enacting the peer-assessor role and what kind of qualifications they deem essential for an assessor. The rest of the questions focused on students' perceptions and attitudes about possible benefits that could emerge when endorsing the peer-assessor and peer-assessee role. The same questions were posed to students during the post-instructional interviews, in addition to questions related to their experience of enacting peer-assessment in this study. Interview data could give us insights into students' perceptions towards peer-assessment.

Finally, students' learning products were also stored on the server that hosted the SCY-Lab but also the Dropbox platform, along with the corresponding log files which included information necessary for our analysis (e.g., first version of each learning product, next versions resulted after revisions etc.). The content of the learning products provided a measure of students' learning achievements and progress.

Data Analysis

We used a mixed-methods perspective that involved both qualitative and quantitative approaches to analysing data. In particular, data were firstly analysed qualitatively and then also treated quantitatively with the use of the SPSS™ software, except the interviews which were merely analysed with open coding for triangulation purposes.

In order to develop a sense of students' typical responses and actions while carrying out the roles of peer assessor and assessee, we used the screen captured data collected, after isolating the episodes in which students enacted these roles. We then coded students' actions, which served as the coding unit, for each role separately. The codes emerged through open coding; after the list of codes was finalized (the finalized coding schemes are given in Tables 1 and 2 respectively), all coding was reviewed once more for consistency purposes. To estimate the inter-rater reliability, a second coder who had not participated in the first round of coding repeated the coding process for 20% of the data of each role separately. Cohen's Kappa was calculated at 0.90 for the coding of the peer assessors' and peer assessees' responses and actions. Differences in the assigned codes were resolved through discussion.

The resulting codes were displayed in time–line graphs following the approach of Schoenfeld (1989) (see Figures 2 and 3). We compared the resulting graphs for each role separately, to identify similarities and differences in the combinations of codes over time. The goal was to reveal interrelationships of the coding over time and identify different patterns that might indicate the range of profiles for peer assessors and peer assessees. To attain this, we tried to detect the codes that discriminated between different strategies and we clustered together patterns of actions that were similar. It should be noted that variations concerning the frequency and order of the same codes within a combination were not considered when clustering the combinations (e.g., it did not make a difference if one peer assessee read first peer feedback and then teacher feedback and another peer assessee group did it the opposite way). The goal was to reveal interrelationships of the coding over time and identify different patterns that might indicate different profiles for each role. The video analysis for the episodes of peer assessors and that for the peer assessees were conducted independently from one another. Peer assessors' actions were categorized into two profiles which differed regarding codes 2 and 11. Peer assessees' actions were categorized into three profiles which differed regarding codes 4 and 6. Figures 2 and 3 respectively illustrate a typical example from each profile in either the assessor or assessee role; the x axis in each graph corresponds to time measured in seconds and the y axis corresponds to assessors' (Figure 2) and assessees' actions (Figure 3).

Table 1. The codes that emerged through open coding of the collected episodes corresponding to the assessors' actions

Code	Peer-assessor actions/responses
1	Opening peer's web-portfolio.
2	Opening one's own web-portfolio.
3	Opening the rubric provided.
4	Reading the instructions on how to use the rubric.
5	Examining the rubric's assessment criteria.
6	Reading the criteria of a specific learning artefact (when time devoted is more than 10 seconds).
7	Assigning grade in criterion/a for one learning artefact or the web-portfolio as a whole (this applies for the last 4 criteria of the rubric).
8	Providing comment in criterion/a for one learning artefact or the web-portfolio as a whole (this applies for the last 4 criteria of the rubric).
9	Proposing changes/providing guidance in criterion/a for one learning artefact or the web-portfolio as a whole (this applies for the last 4 criteria of the rubric).
10	Studying a specific learning artefact from the assessee's web-portfolio.
11	Studying one's own learning artefact, same as the one that is being assessed at the moment.
12	Assigning scores to criteria in a non-linear way.
13	Revising a score that has already been assigned to a specific criterion.
14	Revising a comment that has already been assigned to a specific criterion.
15	Revising proposed changes that have already been assigned to a specific criterion.
16	Providing a comment criterion/a for one learning artefact and at the same time reviewing assessee's learning artefact that is being assessed.
17	Suggesting changes with respect to the criterion/a for one learning artefact and at the same time reviewing assessee's learning artefact that is being assessed.
18	Removing/copying a comment (either positive or negative) that has been already given to one criterion to another one.
19	Removing/copying proposed changes that have been already given to one criterion to another one.
20	Adding a new criterion, at the end of the given rubric, along with a score and a comment.
21	Saving fulfilled rubric with added feedback comments and scores.
22	Uploading the fulfilled rubric in one's own web-portfolio located in Drop box.
23	Uploading the fulfilled rubric in assessee's web-portfolio located in Drop box.
24	Irrelevant tasks (e.g., checking emails).

Table 2. The codes that emerged through open coding of the collected episodes corresponding to the assessees' actions

Code	Peer-assessee actions/responses
1	Opening SCY-Lab learning environment.
2	Opening Dropbox platform.
3	Opening teacher feedback (rubric fulfilled by the teacher).
4	Opening peer feedback (rubric fulfilled by a peer from another group).
5	Opening one's own web-portfolio.
6	Studying peer-feedback.
7	Studyingteacher-feedback.
8	Studyingteacher-feedback which has been given to another peer.
9	Opening one's own learning artefact in the SCY-Lab learning environment.
10	Viewing a specific learning artefact in one's own web-portfolio.
11	Viewing a specific learning artefact in the peer's web-portfolio which the student had previously assessed.
12	Viewing a specific learning artefact in another peer's web-portfolio which the student did not assess.
13	Searching information in the web in regards to freshwater ecosystems.
14	Studying background information available in the SCY-Lab learning environment.
15	Revising one's own specific learning artefact.
16	Creating a new learning artefact.
17	Storing a revised learning artefact from the SCY-Lab learning environment to one's own web-portfolio.
18	Saving one's own revised web-portfolio in the Dropbox platform.
19	Quitting the SCY-Lab platform.
20	Quitting the Dropbox platform.

After having identified students' strategy patterns, we analyzed the rest of the data, which supplemented the creation of students' profiles in each role. Therefore a complete peer assessor profile could correspond to a detailed picture of students' actions while assessing the quality of peer feedback produced, prior knowledge in this topic, learning achievements and progress as depicted in the learning objects and perceptions and attitudes towards peer assessment. The peer-assessee profiles were likewise created, following the same rationale.

The peer-feedback analysis focused on qualitative aspects because it has been emphasized in prior research that qualitative feedback is more important than quantitative (e.g., Topping, et al., 2000). Therefore, we merely analyzed the peer-feedback comments provided by assessors in the given rubrics. For analyzing the qualitative peer feedback, we first followed a pre-defined coding scheme for the written comments about judgements (no judgements, positive judgements, negative/critical judgements, proposed changes/suggestions/ guidance offered for the next steps to be taken, justification of judgements). For example, positive judgements concern proper or correct handling of aspects related to the work/artefacts (e.g., web-portfolio) produced by a student (e.g., inclusion of proper material, inclusion of scientifically accurate information), whereas negative judgements concern incorrect or incomplete handling of aspects related to a student's work/artefacts (for more details, see Tseng and Tsai 2007; Chen et al., 2009). Moreover, we coded each judgement involving science content for its scientific accuracy and also for its quality of the learning artefact being assessed. The peer feedback comments were operationalized in terms of the total and average number of (a) positive and negative judgements per peer assessor's feedback, (b) changes/revisions proposed by each peer assessor, (c) justified judgements offered by each peer assessor, (d) valid judgements offered by each peer assessor, (e) peer-feedback comments offered by each peer assessor to each assessee and (f) peer-feedback comments received by each peer assessee from different peer- assessors. The resulted analysis was used further for calculating descriptive statistics and running non-parametric tests analyses.

We analyzed the pre- and post-instructional tests and interviews with open coding following grounded research methodology (Strauss and Corbin, 1998). In the case of the pre- and post-instructional tests, we firstly input students' responses in a spreadsheet and then started clustering the common responses into codes. We then proceeded to axial coding by grading the coded responses and calculating a score for each response in the pre- and post- instructional test respectively. We further isolated and calculated the frequency of students' misconceptions, which were embedded in their responses (an example of one common misconception among the students is given in the Appendix, along with a typical reponse on the same task given in the post-instructional test). The scores that emerged from the qualitative analysis were used later on for non-parametric tests. The analysis of interview data was used for triangulation purposes but also for identifying students' perceptions towards reciprocal peer assessment.

Finally, for independently analyzing the quality of students' learning artefacts or products, we used a modified and adapted coding scheme — developed for the purposes of the SCY project —that focused on the scientific structure and content of the artefacts. A different coding scheme was used for each learning product, based on the targeted learning goals of each product. Based on the coding of each learning product, a score was calculated that reflected the level of attainment of each student with respect to the goals of each learning task. In Table 3 we provide an example of how one learning product (hypotheses) was coded and graded. The scores that emerged from the qualitative analysis —of each type of learning product —were used later for non-parametric tests.

Table 3. Example from the coding scheme that was used for analysing students' learning products

Learning product: Hypotheses		
Assessed aspect of the learning product	Choose what applies	Points assigned (max points= 11)
a) the formulation	a question is formulated, not a claim	0
	a sentence that is not related to the research question	1
	a static sentence related to the research question	2
	a dynamic sentence related to the research question	3
b) dependent variable	missing	0
	not related to the research question	1
	is related to the research question, but the formulation is non-specific/non-measurable	2
	related to the research question and specific, measurable	3
c) independent variable	missing	0
	not related to the research question	1
	is related to the research question, but the formulation is non-specific/non-measurable	2
	related to the research question and specific, measurable	3
d) elaboration	missing	0
	the hypothesis is further explained but the explanation given is not valid	1
	the hypothesis is further explained and the explanation given is valid with relation with background information	2

Inter-rater reliability values (Cohen's Kappa), for all the aforementioned coding processes were found to exceed 0.89 in every case. All data (peer feedback, test scores, quality of learning products) that were exposed to qualitative analysis were also coded and treated quantitatively through the use of non-parametric tests (Kendall's tau-b correlations, Mann-Whitney test, and Kruskal Wallis test).

RESULTS

Analysis of the pre-instructional interviews' data revealed that students had experienced reciprocal peer-assessment before this intervention, but not SSRPA as implemented in this study. In particular students mentioned that in previous courses at the university that were engaged in oral and written peer-assessment activities, without the provision of any particular scaffolds (e.g., rubrics, assessment criteria). In either case it was found that students had positive attitudes towards peer-assessment, as they could see the potential of benefiting from this process in terms of receiving peer-feedback which could be crucial for improving their ongoing work.

The Role of Peer Assessor

What Are the Various Peer-Assessor Action Patterns and Profiles in SSRPA?

Our analysis of the screen-video-captured data revealed two distinct patterns of peer assessors' typical responses and actions or assessor profiles while giving peer feedback: (1) *autonomous assessors* (profile 1) characterizes those students (N = 19) who gave feedback to peers without seeking any additional information from other resources besides the assessment criteria which were given to them in the form of rubrics and (2) *informed assessors* (profile 2) identifies those students (N = 5) who were seeking information from other resources before providing feedback to peers or seeking a benchmark or some reference level against which to provide comments (see Graphs 1 and 2 in Figure 2).

Graph 1 in Figure 2 represents a typical example of the autonomous assessors' actions. At the beginning of the lesson, this student is reading the given instructions on how to use the rubric and then s/he opens the peer-assessees' web-portfolio which was to her/him (codes 1, 3 and 4). Then the student starts gives assigning a score and a comment to each criterion of the rubric (codes 7-18; see Table 1 for the exact correspondence of those codes to actions). After having addressed all the criteria, the student saves the whole work and uploads the filled in rubric in the assessees' personal folder located in Dropbox platform (codes 23 and 24). In Graph 1, the codes 2 and 11 — corresponding to actions where students in the assessor role were seeking scaffolds while giving feedback — are missing. Graph 2 in Figure 2 represents a typical example of the informed assessors' actions in which the same overall strategy appeared as in the case of the aforementioned example in Graph 1; however, during the provision of peer feedback, the student sought some scaffolds; in particular, s/he opened her/his own web-portfolio (code 2) and reviewed hers/his own learning artefacts included in the web-portfolio (code 11. The remaining actions were almost the same as those in Graph 1.

The results from the post-instructional interviews shed light on possible justifications of students' actions. Informed assessors claimed that indeed they used as a point of reference their own learning artefacts located in their web-portfolio, in an effort to provide as accurate and reliable as possible feedback comments. They stated that they felt more confident when comparing their own work with that of their peer's while assessing. Autonomous assessors did not make any reference to such a necessity for further scaffolds while enacting the peer assessor role. Students from both profiles stressed the importance of having criteria while peer-assessing that could potentially lead to fairer judgements and therefore the provision of more valid and accurate feedback comments.

Additionally, we compared the structural components of the peer feedback produced by peer assessors of the two profiles. Firstly when analyzing the validity of peer-feedback, we found that almost of students' comments were valid, that is, they reflected a correct diagnosis of either strength or weakness of the models. In addition Mann Whitney tests revealed no statistically significant differences between the types of feedback provided by students of the two different profiles (i.e., students from both profiles provided positive and negative comments of an almost equal quantity), except proposed changes and suggestions for further revisions (Mann Whitney $Z = -2.281$, $p < 0.05$). In particular, informed assessors provided more suggestions for improvements (Mean = 20.50) than those provided by the autonomous assessors (Mean = 11.83), which were found to meet the criterion of scientific validity.

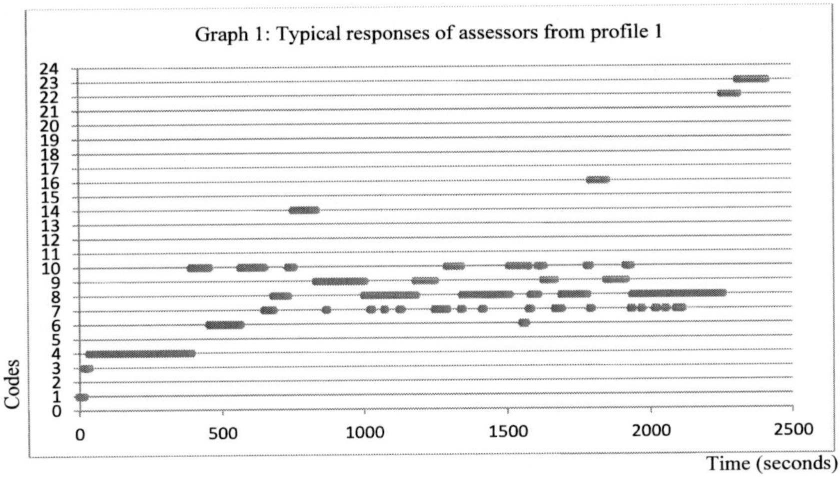

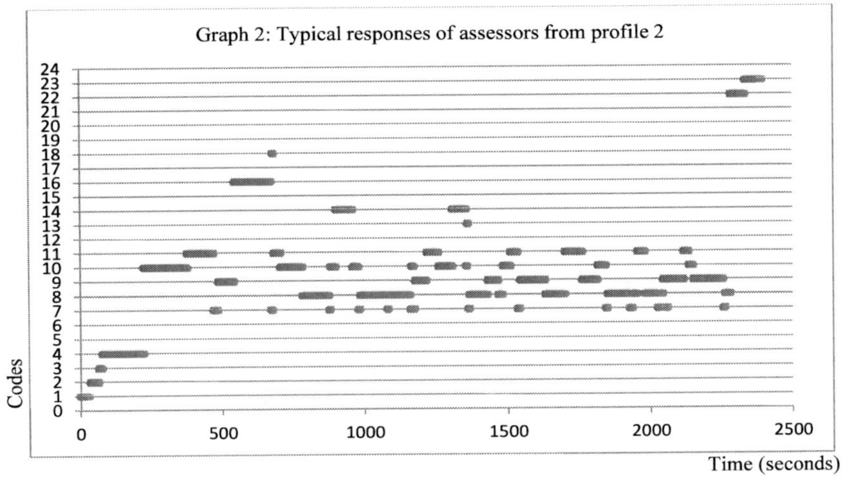

Figure 2. Two representative graphs of peer assessors' actions over time during reciprocal peer assessment. Graph 1 corresponds to the 1st profile (autonomous assessors). Graph 2 corresponds to the 2nd profile (informed assessors). The y-axis represents peer-assessor actions/responses and the x-axis gives the time in seconds. The codes for the y-axis are located in Table 1.

Are the Emerging Peer-Assessor Profiles Associated with Students' Learning Gains in the Chosen Teaching Unit of Ecosystems?

We ran Wilcoxon-rank tests to investigate whether learning gains of students from each of the two profiles; the findings revealed that each group demonstrated positive pre-post learning gains (profile 1: $Z = -3.867$, $p < 0.001$; profile 2: $Z = -2.023$, $p < 0.05$). The quality of their learning artefacts also improved respectively (profile 1: $Z = -2.37$, $p < 0.05$; profile 2: $Z = -3.32$, $p < 0.05$), as measured from the comparison of each learning artefact at its initial version and revised one, for each student separately.

We also compared the performance of autonomous and informed assessors in the pre- and post-instructional tests, as well as the quality of their learning artefacts. The Mann Whitney test results revealed no statistically significant differences among the two profiles regarding their prior knowledge ($Z = -0.649$, $p > 0.05$), their pre-post learning gains ($Z = -0.228$, $p > 0.05$) nor the quality of their learning artefacts ($Z = -0.424$; $p > 0.05$).

The Role of Peer Assessee

What Are the Various Peer-Assessee Action Patterns and Profiles in SSRPA?

Assessees' actions' demonstrated commonalities such as opening the SCY-Lab learning environment at the beginning of the lesson (code 1), opening the Dropbox platform (code 2), opening and reading teacher feedback (codes 3 and 7 respectively), opening one's own learning artefact in the SCY-Lab learning environment (code 9), revising one's own specific learning artefact (code 15) and storing it to one's own web-portfolio (code 17), saving one's own revised web-portfolio in the Dropbox platform (code 18) and quitting the SCY-Lab and Dropbox platform (codes 19 and 20 respectively) and others.

The analysis of the screen-video-captured data revealed three distinct patterns of peer assessees' typical responses and actions while receiving peer feedback (see Figure 3). *Teacher trusting assessees* (profile 1) characterizes those students ($N = 7$) who reviewed only teacher feedback and proceeded with revisions in their own web-portfolios (see Graph 1 in Figure 3: codes 4 and 6 are missing); *teacher trusting and skeptical towards peer feedback assessees* (profile 2) identifies those students ($N = 12$) who reviewed both teacher and peer-feedback but devoted more time on studying teacher feedback (see Table 4; see also Graph 2 in Figure 3: codes 4 and 6 are present, frequency of code 7 is greater than the frequency of code 6). Wilcoxon rank tests revealed that the mean time that *teacher trusting and skeptical towards peer feedback assessees* spent on reviewing teacher and peer feedback respectively varied statistically significant ($Z =- 3.059$; $p < 0.01$), as well as the frequency of revisiting teacher and peer feedback respectively (Z = -3.062; $p < 0.01$) (see Table 4).

Table 4. Comparing the mean time spent on reviewing teacher/peer feedback and the frequency of revisiting teacher/peer feedback for teacher trusting and skeptical towards peer feedback assessees (profile 2)

	Peer feedback	Expert feedback	Wilcoxon rank test Z
Mean time spent on reviewing feedback (in seconds)	233.85	1279.23	-3.059**
Frequency of revisiting feedback	4.23	13.08	-3.062**

** $p < 0.01$.

Those students also proceeded with revisions in their web-portfolios, some of which were proposed by their peers; however *teacher trusting and skeptical towards peer feedback assessees* adopted proposed changes by peers only when those changes were proposed by the teacher as well. In addition, assessees from this profile seemed to follow two different strategies: the first strategy involved a fast check of the peer feedback content at the beginning of the revision phase and then focusing on the teacher feedback content merely; the second strategy involved a thorough check of the teacher feedback during the whole revision phase and an instant check of the peer feedback at certain points. Finally *teacher and peer trusting assessees* (profile 3) (N = 5) reviewed both teacher and peer-feedback and devoted same amount of time on studying them (see Table 5; see also see Graph 3 in Figure 3: codes 4 and 6 are present, frequency of code 7 is almost equal to frequency of code 6). Wilcoxon rank tests revealed that the mean time that *teacher and peer trusting assessees* spent on reviewing teacher and peer feedback did not vary statistically significant (Z = -0.944; $p > 0.05$); the same appeared to apply with respect to the frequency of revisiting teacher and peer feedback respectively (Z = -0.736; $p > 0.05$).

They cross-checked critical comments (negative judgments and proposed changes) received from peer and teacher feedback and revised their web-portfolio based on the common critical comments.

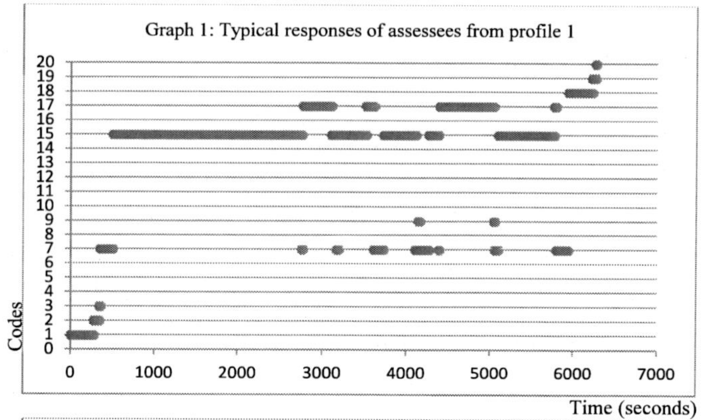

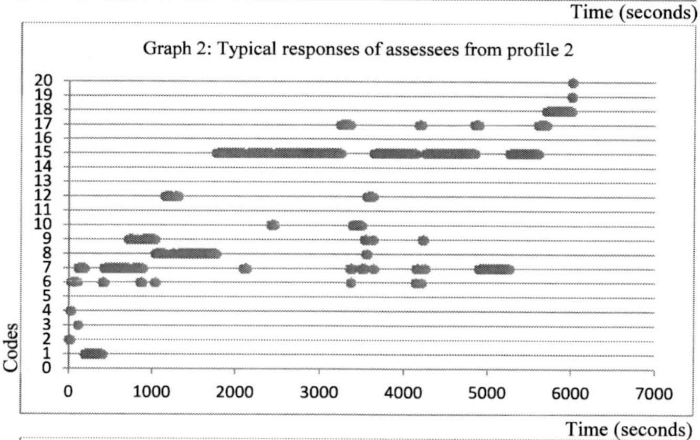

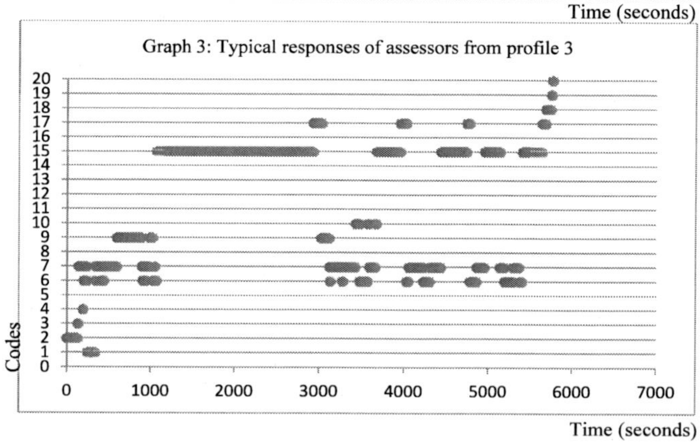

Figure 3. Two representative graphs of peer assessees' actions over time during reciprocal peer assessment. Graph 1 corresponds to the 1st profile (teacher trusting assessees), graph 2 corresponds to the 2nd profile (teacher trusting and skeptical towards peer feedback assessees) and graph 3 corresponds to the 3rd profile (teacher and peer trusting assessees). The y-axis represents peer-assessee actions/responses and the x-axis gives the time in seconds. The codes for the y-axis are located in Table 2.

Table 5. Comparing the mean time spent on reviewing teacher/peer feedback and the frequency of revisiting teacher/peer feedback for teacher and peer trusting assessees (profile 3)

	Peer feedback	Expert feedback	Wilcoxon rank test Z
Mean time spent on reviewing feedback (in seconds)	1392.33	1831.67	-0.944^{ns}
Frequency of revisiting feedback	23.54	25.76	-0.736^{ns}

ns = non significant.

Data from the post-instructional interviews provided insights into students' actions. *Teacher trusting assessees* stated that due to time pressure, they decided to rely merely on teacher feedback for revising their web-portfolio. *Teacher trusting and skeptical towards peer feedback assessees* stated among others that even though they reviewed both peer and teacher feedback, they devoted more time studying teacher feedback, since the teacher offered more guidance/suggestions for improvements on how to proceed with specific revisions on their web-portfolio. The analysis of feedback data confirmed this claim, since it was found that indeed the teacher did offer more guidance in comparison to the guidance offered by peer-assessors (see Table 6). In addition those students mentioned in the post-instructional interviews that teacher-feedback was much more precise and critical compared to peer-feedback. The following excerpts are particularly illustrative:

> "...peer feedback confused me in some cases; therefore I chose to focus on the teacher feedback" (assessee's code: M13a); "due to time restrictions and since teacher feedback was more precise I preferred to focus on the teacher feedback and just scan, go through very fast the peer feedback" (assessee's code: M03b); "If I knew who was the peer that assessed my web-portfolio I could make better decisions on whether I should take it into account or not; since it was anonymous and I did not know whether it was good idea to trust my peer, I preferred to focus on teacher feedback" (assessee's code: M02a).

Finally *teacher and peer trusting assessees* mentioned that peer and teacher feedback resembled; that could be a reason to why those students seemed to appreciate both peer and teacher feedback.

Are the Emerging Assessee Profiles Associated with Students' Learning Gains in the Chosen Teaching Unit of Ecosystems?

Despite the different actions followed, assessees from all three profiles proceeded with almost equal number of revisions on their web-portfolios, which did not diverge statistically significantly (mean number of revisions: profile 1 = 46; profiles 2 and 3 = 51). Students revised their web-portfolios mainly for the scientific content of their learning artefacts (i.e., they ignored comments about organizational and technical issues of their web-portfolio).

We used the Kruskal-Wallis test to compare among the three profiles students' performance in the pre- and post-instructional tests, the misconceptions that emerged from the open coding process of the pre- and post-instructional tests, as well as, the improvement of the quality of their learning artefacts. The results revealed no statistically significant differences among the three profiles regarding the improvement of the quality of their learning artefacts ($H = 0.662$, $p > 0.05$), their learning gains as measured by the post-pre instructional tests scores' difference ($H = 0.384$, $p > 0.05$) nor their pre-and post- misconceptions as revealed through the tests analysis ($H = 0.158$, $p > 0.05$).

Moreover, Kruskal-Wallis tests revealed that no statistically significant differences were present in the structural components of peer-feedback received among students of the three profiles. In an effort to understand why the students acted in a different way while receiving peer and teacher feedback, we first compared the negative, positive comments and guidance offered from peer and teacher feedback in the case of *teacher trusting and skeptical towards peer feedback assessees* (profile 2). The Wilcoxon rank test revealed that, peer and teacher feedback differed with respect to all the aforementioned variables. Particularly, those assessees received more negative and positive comments from their peers than from the teacher; they also received more guidance on how to proceed with revisions from the teacher rather than from their peers, which seemed to be a crucial factor during the revision phase (see Table 6).

Likewise, we compared the structural components of the teacher and peer feedback delivered to *teacher and peer trusting assessees* (profile 3). The findings suggest that peer and teacher feedback diverged only in regards to the number of negative comments provided and interestingly *teacher and peer trusting assessees* received an almost equal number of proposed changes from their peers and the teacher (see Table 7).

Table 6. Comparing peer and teacher feedback delivered to teacher trusting and skeptical towards peer feedback assessees (profile 2)

	Teacher feedback	Peer feedback	Wilcoxon rank test Z
Negative judgments	138	220	-1.84*
Positive judgments	434	581	-2.51*
Proposed changes	275	166	-3.064**

* p < 0,05; ** p < 0,01.

Table 7. Comparing peer and teacher feedback delivered to teacher and peer trusting assessees (profile 3)

	Teacher feedback	Peer feedback	Wilcoxon rank test Z
Negative judgments	45	151	-2.023*
Positive judgments	213	222	-0.271ns
Proposed changes	117	80	-0.944ns

ns = non significant; * p < 0,05.

DISCUSSION

This study focused on examining students' strategy patterns while enacting reciprocal peer assessment– resorting to skills such as judging the performance of a peer (Sluijmans, 2002) in order to provide feedback — and while receiving feedback from their peers and a teacher. The study revealed that student teachers who enacted the peer-assessor role — in an inquiry-oriented learning environment after having studied teaching-learning material about 'marine ecosystems' — seemed to display two different strategy action profiles. They either provided feedback relying on their own and the assessment criteria which were offered to them in the form of rubrics (*autonomous* assessors) or they revisited their own learning products as support for judging their peers' learning products, apart from the assessment criteria (*informed* assessors). Similar findings have been also reported previously (Hovardas et al., 2014; Tsivitanidou and Constantinou, 2016). In the study of Hovardas et al. (2014) secondary school peer assessors relied on their own learning artefacts before providing peer feedback, even when they had been given assessment criteria in the form of rubrics, likewise in the present study. The same discrimination of assessors' heuristics and strategy patterns has also been found in Tsivitanidou and Constantinou (2016) study, even though assessors in that case were not supported with the provision of assessment criteria as in the present study. We endorse the same terminology for labeling the emerged profiles (*autonomous* and *informed* assessors) as in both studies assessors were found to follow comparable strategies, independently of the scaffolds that were offered to them while peer-assessing.

It emerged from the post-instructional interviews that *informed* assessors in this study expressed self-doubt about their ability to provide feedback of adequate quality. Even though students were already familiar with the peer-assessment method, since they had experienced it before even in another context and settlement, as confirmed by the pre-instructional interviews, still the heuristics used by the informed assessors indicate that they had not adequally mastered the assessment skill 'judging the performance of a peer' (Sluijsmans, 2002) or at least, they felt like they had not mastered this skill. Therefore this strategy of cross-checking the content of their own learning artefacts, before providing feedback to their peers could be interpreted as an effort to ensure the accuracy and content validity of their own feedback. Also this strategy resembles the Comparative Judgement (CJ) approach with the provision of a benchmark or point of reference against which they can assess and are more likely to provide valid and reliable feedback (Jones and Alcock, 2014). The CJ approach is based on Thurstone's *"Law of Comparative Judgement"* (Thurstone, 1927, as cited in Jones and Alcock, 2014) according to which people are more reliable in comparing one thing with another than when assigning scores to things (Jones and Alcock, 2012). Previous research has reported similar findings, even though in a different peer-assessment context in which no scaffolds were offered to assessors (Tsivitanidou and Constantinou, 2016). This strategy generates interest, since the CJ approach is not a commonly used method in peer-assessment; nevertheless, it seems that there is potential for CJ approach to be employed by peer-assessors, since informed assessors, as revealed in this study endorsed this approach unconsciously. Future research could further investigate the enactment of the CJ approach in peer-assessment activities, safeguarding that the assessors have gained experienced and/or have received training on peer-assessing.

Interestingly, the findings of this study revealed that, even though *informed* assessors felt self-doubt about their ability to provide feedback, the quality of peer-feedback they produced could be perceived as greater, since they managed to offer more guidance to their peers on the next steps to be taken for improving their work. Researchers have argued that critical comments and suggestions for possible points for improvements (Strijbos and Sluijsmans, 2010) are crucial factors for constructive feedback. Research has shown that peer feedback is constructive if it contains structural components such as suggestions for improvements (Chen et al., 2009; Hovardas et al., 2014; Pare and Joordens, 2008; Ploegh, et al., 2009; Sluijsmans, et al., 2002; Tseng and Tsai, 2007; Tsivitanidou et al., 2011). Therefore is important to create the circumstances under which some students might propose more suggestions for the next steps to be taken by assessees in their future learning. For instance, in the case of web-based learning environments, prompts and hints could encourage students to review other peers' work or to study extra sources of information, before providing feedback, if this could be seen as a strategy for expanding their ideas which could lead to the provision of more suggestions for improvements to peers. Conceding that *informed* assessors offered more constructive feedback to their peers, what we still need to address is how to support students overcoming their self-doubts and low self-confidence of their expertise as assessors. This could be achieved by offering opportunities for a continuous involvement of students in peer-assessment activities and/or the need for training, so as to develop and enhance students' assessment skills (Sluijsmans, 2002).

Moreover, the use of different heuristics, as identified in this study, did not appear to affect in any way the quality of their own learning artefacts nor their learning gains as measured from the variance of the pre- and post-instructional test. The fact that students of these two profiles did not significantly differ in their pre-instructional test scores too, appeared to imply that prior knowledge is not a determinant of the heuristics used. Those findings have been confirmed from previous research findings as well (Tsivitanidou and Constantinou, 2016) which fortifies the claim that heuristics might indeed be self-regulatory actions that are not directly related to students' prior knowledge on a topic, but mostly by the skills required for the task executed, which in this case was peer-assessment. Nonetheless those findings should be replicated in other topics in science learning. The findings on undergraduate students' heuristics and strategy patterns in the peer-assessor role in web-based peer assessment for science learning as summarized in Figure 4.

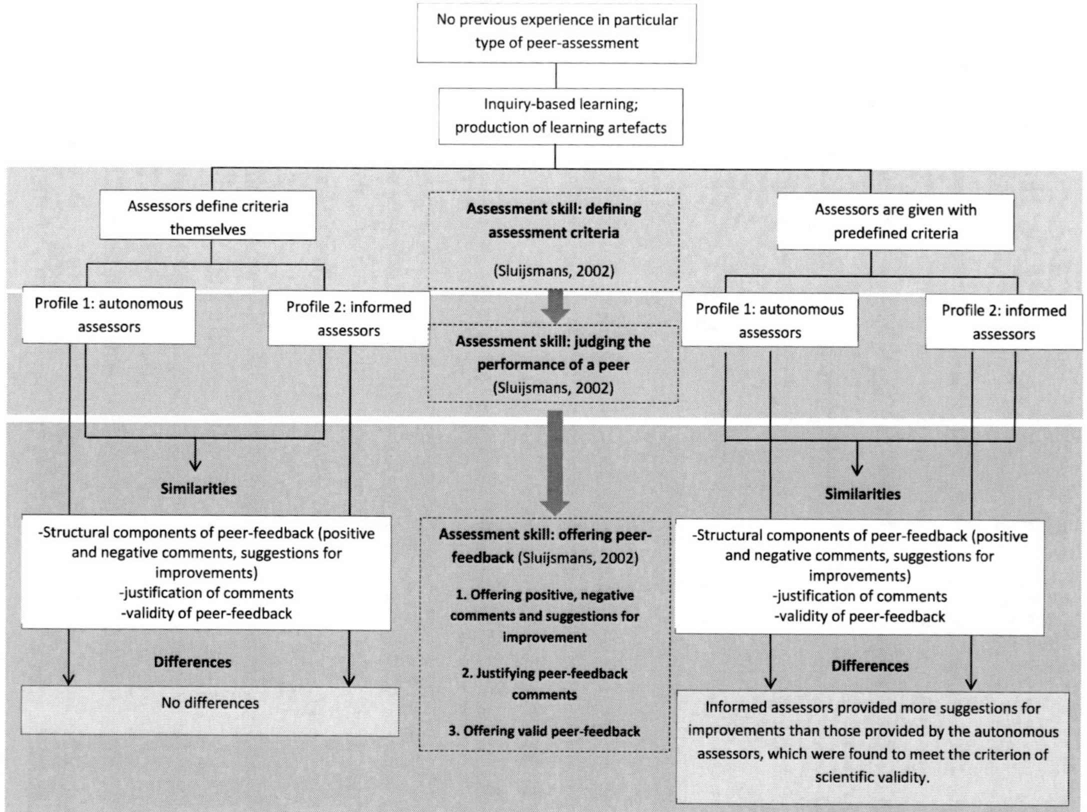

Figure 4. Synopsis of findings on undergraduate students' heuristics and strategy patterns in the peer-assessor role in web-based peer assessment for science learning (Tsivitanidou and Constantinou, 2016; present study).

In the peer-assessee role, the findings of this study revealed three different profiles. Undergraduate teacher students in the present study seem to have followed different heuristics in the peer-assessee role, as those endorsed by secondary school students, reported in Hovardas et al. (2014) study. In the first profile of the present study, students reviewed only teacher feedback and ignored peer feedback received; thus they could be categorized as *teacher trusting* assessees. It could be seen as a paradox, since those students had acknowledged during the pre-instructional interviews, potential benefits on their learning from receiving peer-feedback. They justified their actions later on during the post-instructional interviews, by reporting the time pressure the felt for revising and submitting the final version of their web-portfolio, for the purposes of the course. A time pressure thus, seems to be a crucial factor for deciding on which feedback to rely on: teacher or peer feedback? This factor has been reported previously as a barrier in peer-assessment activities (Hanrahan and Isaacs, 2001; Lin, Liu and Yuan, 2001). It has been also reported previously that students tend to be skeptical towards peer-feedback and prefer to rely on teacher feedback (Peterson and Irving, 2008; Bryant and Carless, 2010; Tsui and Ng, 2000). Results of other studies have shown that peer assessors are rarely regarded as "authority of knowledge" by their peers, leading thus to a feeling of reluctance from assessees to accept and appreciate peer-feedback (Hanrahan and Isaacs, 2001; Strijbos, Narciss and Düunnebier, 2010). Overall previous research has shown that students tend to trust an expert or their teacher more when it comes to choose between expert/teacher feedback and peer feedback (Peterson and Irving, 2008; Bryant and Carless, 2010). The findings of the present study confirm that tendency.

Students of the second profile were those who reviewed both teacher and peer-feedback but devoted more time on studying teacher feedback; hence they could be categorized as *teacher trusting and skeptical towards peer feedback* assessees. They relied mostly on teacher feedback for revising their web-portfolios; however they also adopted proposed changes by peers only when those changes were proposed by the teacher as well. This could be explained by the fact that those students received more guidance statements from the teacher in comparison to the amount of such statements from their peers. Also the fact that those assessee received more positive comments from their peers compared to the teacher feedback seems to have acted as a barrier towards valuing peer feedback; such a finding has been identified in previous research (Tsivitanidou et al., 2011) even though other studies have shown that positive judgments lead to a constructive effect on assessees' intention to revise their work in online peer assessment environments (Tsai and Liang,

2009; Tseng and Tsai, 2007). Altogether the quality of feedback received seems to be a factor which might affect assessees' actions.

Thus far, *teacher trusting and skeptical towards peer feedback* assessees employed strategies resembling those employed by *teacher trusting* assessees. Students from the second profile tended to cross-check suggestions from improvements from their peers to those delivered by the teacher. Time-pressure was revealed again as a restrictive factor for exploiting to a greater extent peer-feedback, as well as their overall attitudes towards peer feedback juxtaposed to teacher feedback. Considering this, it is vital that teachers devote enough time when engaging students in peer-assessment activities, to reduce the time-pressure effect that might be a hampering from reviewing the peer-feedback received, but also the assessor from producing peer-feedback of high quality. Interestingly few students stated during the post-instructional interviews that anonymity of peer-feedback prevented them from relying on it. Alternatively, they would prefer to know the identity of the peer who assessed their work, in order to be able to decide whether to rely on the peer feedback received or not. These findings confront previous findings, according to which anonymity, facilitated often through online systems, can promote positive attitudes of students towards peer assessment and peer feedback (Davies, 2000; 2003; Wen and Tsai, 2006; Tseng and Tsai, 2007; Xiao and Lucking, 2008; Kollar and Fischer, 2010). Perhaps the reason to that paradox may rely on the different perspectives from which anonymity can be confronted. In other words, anonymity might be endorsed by students while enacting the peer-assessor role, but discarded when enacting the peer-assessee role. This could be further examined in future studies.

Finally students of the third profile were those who reviewed both teacher and peer-feedback and devoted same amount of time on teacher and peer-feedback; so-called as *teacher and peer trusting* assessees. Only this profile resembles profile 3 of assessees, as reported in Hovardas et al. (2014) study, in which students proceesed with revisions in their learning artefacts, taking into consideration both teacher and peer feedback. In the present study, teacher and peer feedback received by those assessees were similar in terms of the amount of critical judgements and guidance offered towards the next steps to be taken by the assessees. Furthermore, *teacher and peer trusting* assessees valued peer feedback as confirmed by the post-instructional interviews. This finding stresses again the potential influence of students' attitudes on their actions while enacting peer-assessment, which has been reported previously (Tsivitanidou and Constantinou, 2016). Finally the findings of this study have shown that assessees proceeded to revisions of their work that were suggested from both the teacher and their peers. It seems that those assessees employed triangulation strategies while processing peer and teacher feedback; the fact that assessee groups appear to employ decision-making strategies to screen and process peer and teacher feedback has been also reported in the study of Hovardas et al. (2014).

Finally assessees from all three profiles had the same learning gains at the end of the intervention, as measured through the post-instructional tests but also the quality of their learning artefacts. That is particularly interesting since the heuristics used and the feedback upon which they relied while revising their learning artefacts, did not appear to affect in any way the scientific quality and accuracy of their work neither their content knowledge. In other words, assessees who relied only on teacher feedback (profile 1), confronted as "authority knowledge" for them, had the same learning gains and delivered learning artefacts of the same quality as the students who relied on both peer and teacher feedback (profile 3). However the cross-checking strategies employed by the *teacher and peer trusting* assessees, infers the stimulation of their reflection and in-depth thinking, which might lead to the enhancement of higher order cognitive skills. This needs to be further examined, with the ultimate goal to better understand how different strategies and heuristics used by students, might impact their cognitive and meta-cognitive development. The findings on undergraduate students' heuristics and strategy patterns in the peer-assessee role in web-based peer assessment for science learning are summarized in Figure 5.

Conclusion and Practical Implications

Given the rather small sample size (N = 24) and specific subject pool investigated (undergraduate student teachers in primary education), the findings of this study should be interpreted with caution. Studies with bigger sample sizes and other sample settings should be conducted to examine if these results can be replicated. In addition, future research could focus on the comparative judgement approach and to further examine its potential effectiveness in the peer assessment process, because, as our study has shown, some students tended to use this approach unconsciously (without being instructed to do so) for judging their peer's work.

Overall, the findings of this study can have implications for researchers and educators in higher education about the needs of students engaged in

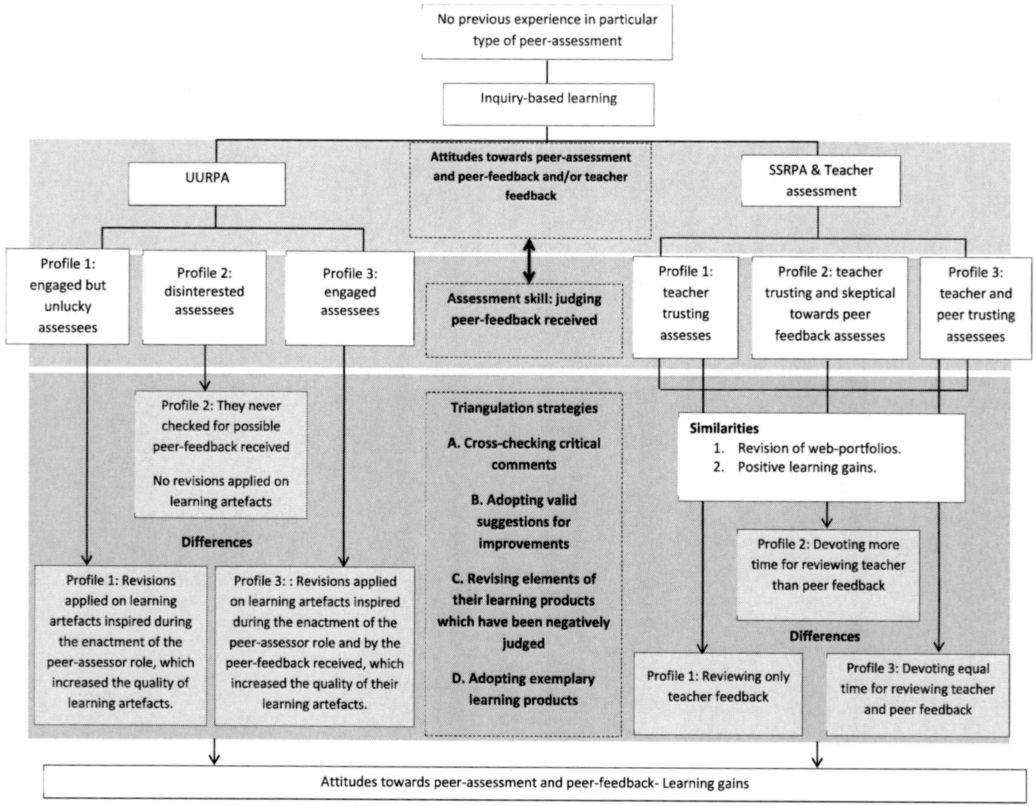

Figure 5. Synopsis of findings on undergraduate students' heuristics and strategy patterns in the peer-assessee role in web-based peer assessment for science learning (Tsivitanidou and Constantinou, 2016; present study). UURPA stands for Unsupported and Unstructured Reciprocal Peer Assessment; SSRPA stands for Supported and Structured Reciprocal Peer Assessment.

implementing peer assessment. Firstly, educators and practitioners could consider the different profiles that seem to appear in the role of the assessor and the assessee respectively, before employing peer-assessment methods in their class or organization, so as to provide the essential degree of scaffolding that might be needed, depending on the context and the individuals involved. This would safeguard a smooth enactment of this activity enhanced learning for the individuals involved. For example, in this study, informed assessors sought for a benchmark or a point of reference against which they could assess their peers' work, resembling the comparative judgement (CJ) approach, even though they were provided with assessment criteria in the form of rubrics. This implies that, even when teachers offer scaffolds (e.g., assessment criteria) to students during the peer-assessment process, students may still need to endorse alternative approaches for of judging their peers' learning artefacts; in this sense teachers should offer more opportunities to students on how to implement peer-assessment.

Designers of web-based learning platforms could utilize this piece of information for developing appropriate technology and tools in which assessees could upload their assignments and the assessors could assess them by means of comparisons with other assignments in relation to the competence under emphasis. Along this way additional scaffolds could be offered in form of hints and prompts for supporting the accomplishment of the peer-assessment activities, in both the roles of assessor and assessee (e.g., examining additional sources of information while enacting the assessor role; cross checking peer-feedback from multiple assessees and filtering the important pieces of information that they might need for their future learning, by also enhancing at the same time their self- reflection and decision making skills while enacting the assessee role).

Furthermore, the findings of this study have revealed that, students' perceptions towards peer and teacher feedback, might affect their actions as assessors and / or assessees; also this has been reported previously (Tsivitanidou and Constantinou, 2016). It is important that teachers eliminate their authority perception and let the students appreciate the value of peer-feedback, even when peer and teacher feedback are offered at the same time. As proposed by Tsivitanidou and Constantinou (2016) it is also important that teachers provide opportunities in which students acknowledge the benefits of peer-assessment; not only when receiving peer-feedback but also when producing peer-feedback and reviewing other peers' work, by elucidating the purpose of this method (assessment, reflection on owns' work, sharing ideas etc.), besides engaging students' in different types of peer-assessment activities. Morevoer, the possible reluctance of students to accept their peers as trusted and competent assessors could be overcomed through involvement and training in peer-assessment and acquisition experience processes. Previous experience or training in peer-assessment, not only supports the role of peer-assessors, for improving the quality of the feedback they produce (Van Steendam, Rijlaarsdam, Sercu and Van den Bergh, 2010), but can also can boost the peer-assessee role, accepting peer- feedback (Gielen, Peeters, Dochy, Onghena and Struyven, 2010). This is crucial, since previous findings have shown that students could actually benefit to a greater extent from the reception of peer-feedback rather than teacher feedback (Yang, Badger and Yu, 2006; Tsivitanidou et al., 2011). Finally it is vital that teachers devote enough time when engaging students in peer-assessment activities, in an effort of reducing the time-pressure effect that might be hampering the assessees from reviewing the peer-feedback received (Tsivitanidou and Constantinou, 2016).

Concluding, undergraduate students might follow the same heuristics and strategies patterns in the peer-assessor role, regardless of the scaffolds offered to them, as the findings of this study replicate the findings on Tsivitanidou and Constantinou (2016) study. However, the educational level may affect the heuristics followed by peer-assessors; since the assessor profiles that emerged in the present study diverge from the profiles of secondary school students anticipated in Hovardas et al. (2014) study. Likewise for the peer-assessee role, students from different educational levels seem to employ different strategies when receiving peer and expert feedback. This calls for further research and in particular in the primary education level, as generates interest to examine how peer-assessors and assessees act in that level. It also suggests that teachers should bear in mind that different strategies and heuristics may be observed by students from different educational levels during peer-assessment activities.

ACKNOWLEDGMENTS

The authors would like to thank Dr Z. Zacharia for facilitating the data collection process and the coordinators and partners of the research project Science Created by You (SCY) for supporting this study. The SCY project was funded by the European Community under the Information and Communication Technologies (ICT) pillar of the 7th Framework Programme for R&D (Grant agreement no.: 212814). The report of this study does not represent the opinions of the SCY project, nor those of the European Community.

References

Bell, B. and Cowie, B. (2001). The characteristics of formative assessment in science education. *Science education, 85*(5), 536-553.

Black, P. and Harrison, C. (2001). Feedback in questioning and marking: The science teacher's role in formative assessment. *School science review, 82*(301), 55-61.

Ballantyne, R., Hughes, K. and Mylonas, A. (2002). Developing procedures for implementing peer assessment in large classes using an action research process. *Assessment and Evaluation in Higher Education, 27*(5), 427–441.

Bouzidi, L. H. and Jaillet, A. (2009). Can online peer assessment be trusted?. *Journal of Educational Technology and Society, 12*(4), 257.

Brown, G., Bull, J. and Pendlebur, M. (1997). *Assessing Student Learning in Higher Education* (London, Routledge).

Bryant, D. A. and Carless, D. R. (2010). Peer assessment in a test-dominated setting: empowering, boring or facilitating examination preparation? *Educational Research in Policy and Practice, 9,* 3-15.

Cestone, C. M., Levine, R. E. and Lane, D. R. (2008). Peer assessment and evaluation in team-based learning. *New Directions for Teaching and Learning, 116,* 69-78.

Chen, N.-S., Wie, C.-W., Wu, K.-T. and Uden, L. (2009). Effects of high level prompts and peer assessment on online learners' reflection levels. *Computers and Education, 52,* 283-291.

Cheng, Y. C. and Ku, H. Y. (2009). An investigation of the effects of reciprocal peer tutoring. *Computers in Human Behavior, 25*(1), 40-49.

Chang, C.-C., Tseng, K.-H., Chou, P.-N. and Chen, Y.-H. (2011). Reliability and validity of web-based portfolio peer assessment: A case study for a senior high school's students taking computer course. *Computers and Education, 57*(1), 1306–1316.

Cho, K. and Schunn, C. D. (2007). Scaffolded writing and rewriting in the discipline. *Computers and Education, 48,* 409-426.

Crane, L. and Winterbottom, M. (2008). Plants and photosynthesis: peer assessment to help students learn. *Journal of Biological Education, 42,* 150-156.

Davies, P. (2006). Peer assessment: judging the quality of students' work by comments rather than marks. *Innovations in Education and Teaching International, 43,* 69–82.

De Jong, T., Van Joolingen,W., Giemza, A., Girault, I., Hoppe, U., Kindermann, J., et al. (2010). Learning by creating and exchanging objects: the SCY experience. *British Journal of Educational Technology, 41*(6), 909–921.

De Jong, T., Weinberger, A., Van Joolingen, W., Giemza, A., Girault, I., Hoppe, U., et al. (2012). Using scenarios to design complex technology-enhanced learning environments. Education, Technology, *Research and Development, 60,* 883–901.

Dochy, F., Segers, M. and Sluijsmans, D. (1999). The use of self-, peer and co-assessment in higher education: a review. *Studies in Higher Education, 24,* 331-350.

Duran, D. and Monereo, C. (2005). Styles and sequences of cooperative interaction in fixed and reciprocal peer tutoring. *Learning and Instruction, 15*(3), 179-199.

Falchikov, N. (1995). Peer feedback marking: developing peer assessment. *Innovations in Education and Training International, 32,* 175–187.

Falchikov, N. (2003). Involving students in assessment. *Psychology Learning and Teaching*, 3, 102-108.

Falchikov, N. and Goldfinch, J. (2000). Student peer assessment in higher education: a meta-analysis comparing peer and teacher marks. *Review of Educational Research, 70,* 287-322.

Gielen, S., Peeters, E., Dochy, F., Onghena, P. and Struyven, K. (2010). Improving the effectiveness of peer feedback for learning. *Learning and Instruction, 20,* 304-315.

Hafner, J. C. and Hafner, P. M. (2003). Quantitative analysis of the rubric as an assessment tool: an empirical study of student peer-group rating. *International Journal of Science Education, 25,* 1509–1528.

Hanrahan, S. J. and Isaacs, G. (2001). Assessing self- and peer-assessment: the students' views. *Higher Education Research and Development, 20,* 53-70.

Harlen, W. (2007). Holding up a mirror to classroom practice. *Primary Science Review, 100,* 29–31.

Harrison, C. and Harlen, W. (2006). *Children's self- and peer-assessment.* In W. Harlen (Ed.), Ase guide to primary science education (pp. 183–190). Hatfield: Association for Science Education.

Hovardas, T., Tsivitanidou, O. E. and Zacharias, C. Z. (2014). Peer versus Expert feedback: Investigating the quality of peer feedback among secondary school students assessing each other's science web-portfolios. *Computers and Education, 71,* 133-152.

Hwang, C.-J, Hung, C.-M. and Chen, N.-S. (2014). Improving learning achievements, motivations and problem-solving skills through a peer assessment-based game development approach. *Education Tech Research Dev., 62*, 129-145.

Jones, I. and Alcock, L. (2012). *Summative peer assessment of undergraduate calculus using adaptive comparative judgement.* In P. Iannone and A. Simpson. (Eds.). Mapping university mathematics assessment practices (pp 63 – 74). Norwich, U.K.: University of East Anglia.

Jones, I. and Alcock, L. (2014). Peer assessment without assessment criteria. *Studies in Higher Education, 39*(10), 1774-1787.

Kollar, I. and Fischer, F. (2010). Peer assessment as collaborative learning: A cognitive perspective. *Learning and Instruction, 20*(4), 344-348.

Lin, S. S. J., Liu, E. Z. F. and Yuan, S. M. (2001). Web-based peer assessment: feedback for students with various thinking styles. *Journal of Computer Assisted Learning, 17,* 420-432.

Orsmond, P., Merry, S. and Reiling, K. (1996). The importance of Marking Criteria in the Use of Peer-Assessment. *Assessment and Evaluation in Higher Education. 21* (3), 239-250.

Orsmond, P., Merry, S. and Reiling, K. (2000). The use of student derived marking criteria in peer and self-assessment. *Assessment and Evaluation in Higher Education, 25,* 23–38.

Paré, D. E. and Joordens, S. (2008). Peering into large lectures: examining peer and expert mark agreement using peerScholar, an online peer assessment tool. *Journal of Computer Assisted Learning, 24*, 526–540.

Pedaste, M., de Jong, T., Sarapuu, T., Piksööt, J., van Joolingen, W. R. and Giemza, A. (2013). Investigating Ecosystems as a Blended Learning Experience. *Science, 340* (6140), 1537-1538.

Pellegrino, J. W., Chudowsky, N. and Glaser, R. (2001). Knowing what students know: The science and design of educational assessment. Washington, DC: National Academic Press.

Peterson, E. R. and Irving, S. E. (2008). Secondary school students' conceptions of assessment and feedback. *Learning and Instruction, 18,* 238-250.

Ploegh, K., Tillema, H. H. and Segers, M. S. R. (2009). In search of quality criteria in peer assessment practises. *Studies in Educational Evaluation, 5,* 102-109.

Pollitt, A. (2012). The method of adaptive comparative judgement. *Assessment in Education: Principles, Policy and Practice, 19*(3), 281–300.

Prins, F. J., Sluijsmans, D. M. A., Kirschner, P. A. and Strijbos, J.-W. (2005). Formative peer assessment in a CSCL environment: a case study. *Assessment and Evaluation in Higher Education, 30,* 417–444.

Searby, M. and Ewers, T. (1997). An evaluation of the use of peer assessment in higher education: a case study in the School of Music, Kingston University. *Assessment and Evaluation in Higher Education, 22,* 371–383.

Schoenfeld, A. H. (1989). Teaching mathematical thinking and problem solving. In L. B. Resnick, and B. L. Klopfer (Eds.), *Towards the thinking curriculum: Current cognitive research* (pp. 83–103). Washington DC: ASCD.

Sluijsmans, D. M. A. (2002). *Student involvement in assessment, the training of peerassessment skills.* Interuniversity Centre for Educational Research.

Sluijsmans, D. M. A., Brand-Gruwel, S. and Van Merriënboer, J. J. G. (2002). Peer assessment training in teacher education: effects on performance and perceptions. *Assessment and Evaluation in Higher Education, 27,* 443-454.

Strauss, A. and Corbin, J. (1998). *Basics of qualitative research: Techniques and procedures for developing grounded theory (2nd ed.).* Thousan Oaks, CA: Sage.

Strijbos, J. W., Narciss, S. and Dünnebier, K. (2010). Peer feedback content and sender's competence level in academic writing revision tasks: are they critical for feedback perceptions and efficiency? *Learning and Instruction, 20*(4), 291-303.

Strijbos, J. W. and Sluijsmans, D. M. (2010). Unraveling peer assessment: methodological, functional and conceptual developments. *Learning and Instruction, 20,* 265–269.

Topping, K. (1998). Peer assessment between students in colleges and universities. *Review of Educational Research, 68(3),* 249-276.

Topping, K. J. (2003). *Self and peer assessment in school and university reliability, validity and utility.* In M. Segers, F. Dochy and E. Cascaller (Eds.), Optimising New Modes of Assessment: In Search of Qualities and Standards (pp. 55–87). The Netherlands: Kluwer Academic Publishers.

Topping, K., Smith, F. F., Swanson, I. and Elliot, A. (2000). Formative peer assessment of academic writing between postgraduate students. *Assessment and Evaluation in Higher Education, 25,* 149-169.

Tsai, C.-C., Lin, S. S. J. and Yuan, S.-M. (2002). Developing science activities through a network peer assessment system. *Computers and Education, 38 (1-3),* 241-252.

Tsai, C.-C. and Liang, J. C. (2009). The development of science activities via on-line peer assessment: the role of scientific epistemological views. *Instructional Science, 37,* 293– 310.

Tseng, S.-C. and Tsai, C.-C. (2007) On-line peer assessment and the role of the peer feedback: A study of high school computer course. *Computers and Education, 49,* 1161–1174.

Tsivitanidou, E., O., Zacharias, Z. and Hovardas, T. (2011). Investigating secondary school students' unmediated peer assessment skills. *Learning and Instruction, 21*(4), 506-519.

Tsivitanidou, O. and Constantinou, C. (2016). A study of students' heuristics and strategy patterns in web-based reciprocal peer assessment for science learning. *The Internet and Higher Education. 12,* 12-22, DOI:10.1016/j.iheduc.2015.11.002.

Tsui, A. B. M. and Ng, M. (2000). Do secondary L2 writers benefit from peer comments? *Journal of Second Language Writing, 9,* 147-170.

Van Gennip, N. A. E., Segers, M. S. R. and Tillema, H. H. (2010). Peer assessment as a collaborative learning activity: the role of interpersonal variables and conceptions. *Learning and Instruction, 20,* 280–290.

Van Steendam, E., Rijlaarsdam, G., Sercu, L. and Van den Bergh, H. (2010). The effect of instruction type and dyadic or individual emulation of higher-order peer feedback in EFL. *Learning and Instruction, 20,* 316–327.

Weaver, W. and Cottrell, H.W. (1986). Peer evaluation: A case study. *Innovative Higher Education, 11*, 25-39.

Xiao, Y. and Lucking, R. (2008). The impact of two types of peer assessment on students' performance and satisfaction within a Wiki environment. *The Internet and Higher Education, 11*(3), 186–193.

Yang, M., Badger, R. and Yu, Z. (2006). A comparative study of peer and teacher feedback in a Chinese EFL writing class. *Journal of Second Language Writing, 15,* 179-200.

Yang, Y-F. (2011). A reciprocal peer review system to support college students' writing. *British Journal of Educational Technology, 42*(4), 687–700.

APPENDIX A.

A. Criteria Included in the Peer Assessment Rubric for Each Science Web- Portfolio

Category	Learning artefacts	Assessment Criteria
Addressing learning artefacts	Concept map	(1) My peers' concept map is original as to its content, since it contains useful information that I had not thought of. (2) The concept map is constituted by pieces of valid information about a freshwater aquatic ecosystem. (3) When reading the concept map, one can understand what an aquatic ecosystem is and what is comprised from. (4) The concept map is well organized. (5) The several elements of the concept map are well connected. (6) The propositions are labelled. (7) The propositions are valid. (8) The concepts/ideas included in the concept map could be easily understood by our classmates.
	Problem formulation	(9) The problem is clearly defined. (10) My peer has adequately justified why the observation of the two children leads to concern for further investigation. (11) Through the formulation of the problem it appears that my peer has used the available sources of information given in the SCY-Lab learning environment, to justify hers/his response.
	Research questions	(12) My peer's research questions are clearly presented, which means that one can understand what s/he wants to investigate. (13) The research questions are formed in one of the following ways: *"Does factor A affects factor B"* or *"What is the relationship among factor A and B?"* (14) My peer's research questions can be tested with the available tool (SCYDynamics) embedded in the SCY-Lab learning environment. (15) The variables of the research questions are testable and measurable. (16) The independent and dependent variables are clearly defined in all my peer's research questions. (17) My peer's research questions are related with hers/his problem formulation.
	Hypothesis	(18) My peer's hypotheses are clearly presented. (19) The hypotheses are formed as follows: *"If... then."* (20) My peer's hypotheses are related with hers/his research questions. (21) The variables of the hypotheses are testable and measurable. (22) The independent and dependent variables are clearly defined in all my peer's hypotheses.
	Experimental design	(23) The experimental process that my peer will follow for answering hers/his research questions is clearly presented, which means it is understandable to me how my peer will manipulate the SCYDynamics tool, in order to answer hers/his research questions. (24) As far as I can understand while assessing my peer's experimental design, my peer has the essential skills for manipulating an experiment. (25) The variables to be tested, measured and controlled are defined. (26) The way in which the data will be treated is presented.

Category	Learning artefacts	Assessment Criteria
	Data from SCYDynamics	(27) The manipulations done in the SCYDynamics tool are sufficient for carrying out inferences for the chosen research questions. (28) For investigating each one of the chosen research questions, my peers has chosen the right variables to be tested, measured and controlled in the SCYDynamics tool (comment for each research question separately). (29) In the graphs presented, the x axis corresponds to time (t), so as to show the change of the dependent variable in time, for different values of the independent variable.
		(30) As far as I can understand while assessing my peer's graphs, my peer has the essential skills for manipulating variables (e.g., same values for the controlled variables; changing only the independent variable at a time; observing the change in the dependent variable via the graph).
	Inferences	(31) As far as I can understand while assessing my peer's inferences, it seems that my peer has the skill to interpret hers/his observations. (32) The answers given to all research questions are sufficiently substantiated with data collected by my peer during hers/his investigation. (33) My peer is contradicting the conclusion inferred by the results of this investigation with hers/his initial hypothesis.
	Problem solution	(34) The solution to the problem initially presented, is clearly presented. (35) My peer's problem solution is scientifically valid. (36) My peer's problem solution is justified efficiently with hers/his inferences presented above. (37) Through the problem solution it appears that my peer has used the available sources of information given in SCY-Lab learning environment, to justify hers/his response. (38) System dynamics of prey-predator relationships are presented as a main argument for justifying the initial observation of the two children from the initial given story. (39) As far as I can understand while assessing my peer's problem solution, it seems that my peer has the ability to explicate hers/his understanding on a topic.
Addressing web-portfolio overall		(40) My peer has included in hers/his web-portfolio all the learning artefacts needed for completing the ECO mission. (41) As far as I can understand while assessing my peer's web-portfolio, it seems that my peer has used the available information sources in the SCY-Lab learning environment effectively while constructing hers/his learning artefacts. (42) The texts enclosed in my peer's learning artefacts are well structured (in clear language and appropriate vocabulary). (43) As far as I can understand while assessing my peer's web-portfolio, it seems that my peer has the skill to clearly present the knowledge acquired.

Note: Each criterion was accompanied by Likert-scale anchors (i.e., (1) unsatisfactory; (2) moderately satisfactory; (3) (fully) satisfactory) and a space for explaining the reasoning behind the grade and for providing feedback.

B. Example of a Common Misconception among Students, Derived from One of the Pre-Instructional Test's Tasks

Task 3: "In a forest habitat there is a pack of 5 wolves and a herd of 10 deer. How will the number of wolves and deer be differentiated in 4, 8, 12 and 16 months?"

- Make a prediction on how the number population of each one the two species (wolves and deer) will evolve in the short-term (in two months), medium-term (6 months) and long-term (1 year). Explain your reasoning.
- Based on your predictions, create a graph where the time is plotted in the x-axis and the number of the two species in the y-axis. Use two different colors: blue for the wolves, red for the deer.

Graph 1

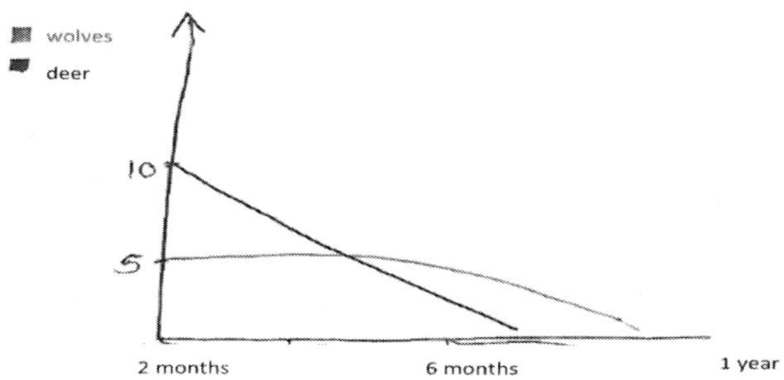

Graph 2

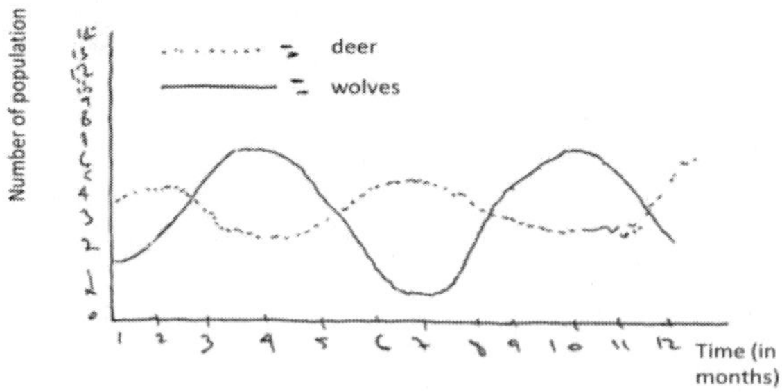

Figure 6. Graph 1 illustrates a typical misconception appeared in students' responses given in the pre-instructional test; Graph 2 illustrates a typical response from students given in the post-instructional test.

BIOGRAPHICAL SKETCHES

Name: Olia E. Tsivitanidou

Affiliation: University of Cyprus, Department of Educational Sciences

Education: Olia Tsivitanidou is a PhD holder in Learning in Natural Sciences at the University of Cyprus (2010-2014). Her PhD thesis is titled: *"Investigating student teachers' profiles assessors and assessees in the context of reciprocal web-based formative peer-assessment, in a learning environment on ecosystems"*. She holds a Bachelor's Degree in Primary School Education from the University of Cyprus (2008) and a Master in Learning in Natural Sciences from the University of Cyprus (2010), and has defended a Thesis entitled: *"Implementation of peer-assessment of digital artifacts, through the use of students' web-portfolios in a science course in secondary education"*. Currently she is working as a post-doctoral researcher at the Department of Educational Sciences at the University of Cyprus (September 2015- present) and she is also studying Physics (B.Sc.) at the University of Cyprus (January 2016-present).

Address: University of Cyprus, P. O. Box 20537, CY 1678,
Nicosia, Cyprus

Research and Professional Experience: She is currently working at the University of Cyprus as a postdoctoral researcher at the Department of Educational Sciences (September 2015-present). Part of her duties involves research on the domain of formative assessment in Science Education and teaching under-graduate teacher students. She is serving as a reviewer in the international research journal *Educational Research Review*. Dr. Tsivitanidou has been employed by the University of Applied Sciences and Arts Northwestern Switzerland FHNW as a post-doctoral researcher (July 2014-August 2015). She has been also a member of the Learning in Science Group (LSG) of the University of Cyprus (Present–July 2008) and being employed as post-graduate researcher and teaching assistant in the Department of Educational Sciences of the University of Cyprus. Olia has worked for various European Founded Research Projects. For example, she has been involved in the ASSIST-ME project which is funded by the European Community under the 7th Framework Programme (2013-2017). ASSIST-ME is a high level research project that investigates formative and summative assessment methods to support and to improve inquiry-based approaches in European science, technology and mathematics education. She has also worked for the research project Science Created by You (SCY), which was funded by the European Community under the Information and Communication Technologies (ICT) theme of the 7th Framework Programme for R&D (Grant agreement no.: 212814). SCY aimed to take Science Education to the next level by developing a flexible, open-ended learning environment that engages and empowers adolescent learners. She has also worked for various National Founded Research Projects [e.g., (i) in the research project *«Checks und Aufgabensammlung»*, which is conducted under the leadership of the Institut für Bildungsevaluation-Universität Zürich (2014-2015); (ii) involvement in a case study under the cooperation of the Pedagogical Institute, the Cardet Research Group of the University of Nicosia and the Learning Research Group in Natural and Environmental Sciences of the University of Cyprus, October-December 2008, on the integration of ICT in Education].

She has also experience in teaching across all grades of the primary educational level, as a Science Teacher employed in The G C School of Careers (Sept 2013- June 2014). Overall her research interest falls within the area of formative assessment and in particular web-based peer-assessment procedures in computer supported learning environments with the aim of identifying their possible effects on students' learning.

She is a member of:

- European Association for Research on Learning and Instruction (EARLI)
- JURE (JUnior REsearchers) network of EARLI
- European Science Education Research Association (ESERA)
- JURE (JUnior REsearchers) network of ESERA

She received the best JURE Award for innovative paper at the 7th EARLI Sig 1: Assessment and Evaluation 2014 Conference held in Madrid, Spain (27-29 August). Paper: *Examining undergraduate teacher assessors' and assesses' heuristics and their behavioural profiles in a web-based peer assessment context in science.*

Publications Last 3 Years:

I. Publications in Scientific Journals with Blind Review System

1. Tsivitanidou, O. and Constantinou, C. (2016). A study of students' heuristics and strategy patterns in web-based reciprocal peer assessment for science learning. *The Internet and Higher Education. 12,* 12-22, DOI:10.1016/j.iheduc.2015.11.002.
2. Hovardas, T., Tsivitanidou, O. E. and Zacharias, C. Z. (2014). Peer versus Expert feedback: Investigating the quality of peer feedback among secondary school students assessing each other's science web-portfolios. *Computers and Education, 71,* 133-152.

II. Publications in Conference Proceedings with Review System

3. Tsivitanidou, O. and Labudde, P. (2016). Can peer-assessment inform teachers about secondary-school students' mastery of the modeling competence? In: Lavonen, J., Juuti, K., Lampiselkä, J., Uitto, A. and Hahl, K. (ed.) Science Education Research: Engaging learners for a sustainable future (Proceedings of ESERA 2015). Helsinki: University of Helsinki.
4. Felchlin I., Tsivitanidou O., Beerenwinkel A. and Moser U. (2016). Simulation and modelling: designing a summative assessment test in Biology. In: Lavonen, J., Juuti, K., Lampiselkä, J., Uitto, A. and Hahl, K. (ed.) Science Education Research: Engaging learners for a sustainable future (Proceedings of ESERA 2015). Helsinki: University of Helsinki.
5. Tsivitanidou, O., Zacharias, Z. and Hovardas, T. (2014). Examining whether secondary school students make changes suggested by expert or peer assessors in the science web-portfolios, In C. P. Constantinou, N. Papadouris and A. Hadjigeorgiou (Eds.), E-Book Proceedings of the ESERA 2013 Conference: Science Education Research For Evidence-based Teaching and Coherence in Learning. Part 11 (co-ed. Robin Millar and Jens Dolin), (pp.68-75) Nicosia, Cyprus: European Science Education Research Association. ISBN: 978-9963-700-77-6.

Name: Constantinos P. Constantinou

Affiliation: University of Cyprus, Department of Educational Sciences

Education: C.P. Constantinou has a PhD in Physics from the University of Cambridge and has worked at Washington State University and the University of Washington.

Address: University of Cyprus, P. O. Box 20537, CY 1678,
Nicosia, Cyprus

Research and Professional Experience: C. P. Constantinou is the Vice Rector for International Relations, Administration and Finance at the University of Cyprus where he is also serving as a Professor in Science Education and Director of the Learning in Science Group. He has a PhD in Physics from the University of Cambridge and has worked at Washington State University and the University of Washington. He is a member of the editorial boards of the *International Journal of Science Education*, the *Educational Research Review* and the journal *Learning, Culture and Social Interaction.* He is serving as a reviewer in other international research journals including *Learning and Instruction* and the *Journal of Research in Science Teaching*. He has been active in international educational research over a period of more than 20 years with research interests that focus on the learning and teaching of science as a process of inquiry and the use of educational technologies as a tool for promoting critical evidence-based thinking. The Learning in Science Group uses the results of this research in the development of online learning environments, such as the STOCHASMOS platform, and research-based teaching-learning sequences to promote conceptual understanding and scientific thinking. Dr Constantinou has co-ordinated a number of projects funded by the European Commission and the Cyprus Research Promotion Foundation. He has participated in the High Level Expert Groups that authored the reports *Science Education for Responsible Citizenship* in 2015 and *Europe needs more Scientists!* in 2004; he was the Rapporteur in a group of experts that carried out the mid-term evaluation of the EC Science in Society program in FP6 and he has participated in an expert panel on the educational outreach programme of the European Space Agency. He is currently President of the European Science Education Research Association (ESERA).

Professional Appointments: Professor of Science and Education, University of Cyprus

Publications Last 3 Years:

I. Edited Books

1. N. Papadouris, A. Hadjigeorgiou and C. P. Constantinou (Eds.), 2015.
   Insights from Research in Science Teaching and Learning
   Springer International Publishing
   Book Series: Contributions from Science Education Research, Vol. 2, 2016, Series Editor: Robin Millar
   ISBN: 978-3-319-20073-6 (Print) 978-3-319-20074-3 (Online)
   DOI: 10.1007/978-3-319-20074-3
   C. P. Constantinou, N. Papadouris and A. Hadjigeorgiou (Eds.), 2014.
   Science Education Research For Evidence-based Teaching and Coherence in Learning. e-Book Proceedings of the ESERA 2013 Conference: Nicosia, Cyprus. European Science Education Research Association.
   ISBN: 978-9963-700-77-6.

II. Publications in Scientific Journals with Blind Review System

2. O. Tsivitanidou and C. P. Constantinou
   A study of students' heuristics and strategy patterns in web-based reciprocal peer assessment for science learning.
   The Internet and Higher Education, 2016, Volume 29, pp. 12–22
   DOI:10.1016/j.iheduc.2015.11.002
   N. Papadouris and C. P. Constantinou
   Investigating middle school students' ability to develop energy as a framework for analyzing simple physical phenomena
   Journal of Research in Science Teaching, 2016, Vol. 53, No. 1, pp. 119–145
   DOI: 10.1002/tea.21248
3. K. Iordanou and C. P. Constantinou
   Supporting Use of Evidence in Argumentation through Practice in Argumentation and Reflection in the context of the Socrates Learning Environment.
   Science Education, 2015, Volume 99, No. 2, pp. 282–311
   DOI 10.1002/sce.21152

4. N. Papadouris and C. P. Constantinou
Pre-service elementary school teachers' ability to account for the operation of simple physical systems using the energy conservation law.
Journal of Science Teacher Education, 2014, Volume 25, pp. 911-933
DOI 10.1007/s10972-014-9407-y
5. Chr. Th. Nicolaou and C. P. Constantinou
Assessment of the modeling competence: A systematic review and synthesis of empirical research.
Educational Research Review, 2014, Volume 13, pp. 52-73
DOI 10.10.16/j.edurev.2014.10.001
6. K. Iordanou and C. P. Constantinou
Developing pre-service teachers' evidence-based argumentation skills on socio-scientific issues.
Learning and Instruction, 2014, Volume 34, December 2014, Pages 42–57, DOI: 10.1016/j.learninstruc.2014.07.004
7. A. Mettas and C. Constantinou
Design and Technology Fairs as mechanisms for familiarizing student teachers with problem-solving practices
Inquiry in Primary Science Education (IPSE), 2014, 1, pp. 32-44.
8. X. Danos, C. P. Constantinou, M. Livitzis, Chr. Avraam
Action research to develop a scheme of work to promote creativity and designerly thinking through play.
Design and Technology Education: An International Journal, 2014, 19(2), pp. 17-29.
9. N. Papadouris and C. P. Constantinou
An Exploratory Investigation of 12-Year-Old Students' Ability to Appreciate Certain Aspects of the Nature of Science through a Specially Designed Approach in the Context of Energy.
International Journal of Science Education, 2014, Vol. 36, No. 5, pp. 755-782. DOI:10.1080/09500693.2013.827816.
10. C. P. Constantinou, A. Raftopoulos, G. Spanoudes and D. Natsopoulos
Preschoolers' Construction of Operational Definitions in Magnetism
Journal of Emergent Science, 2013, Issue 5, Summer 2013, pp. 6-21.

11. L. Louca, D. Tzialli, T. Skoulia and C. P. Constantinou
Developing teaching responsiveness to children's inquiry in science: A case study of professional development for pre-school teachers.
Nordic Studies in Science Education, 2013, Vol. 9, No. 1, pp. 66-81.

III. Chapters in Edited Books

12. N. Papadouris and C. P. Constantinou
Design, Development and Refinement of a Teaching-Learning Sequence on the Electromagnetic Properties of Materials
In D. Psillos and P. Kariotoglou (editors) Iterative Design of Teaching-Learning Sequences: Introducing the Science of Materials in European Schools,
Springer International Publishing, 2015, pp. 331-369
ISBN 978-94-007-7808-5
13. C. P. Constantinou
Science Education Research for Enhancing Classroom Learning
In D. Psillos and P. Kariotoglou (editors) Iterative Design of Teaching-Learning Sequences: Introducing the Science of Materials in European Schools,
Springer International Publishing, 2015, pp. 371-375
ISBN 978-94-007-7808-5
14. A. Hadjigeorgiou, N. Papadouris and C. P. Constantinou
Introduction: Insights from Research in Science Teaching and Learning
In N. Papadouris, A. Hadjigeorgiou and C. P. Constantinou (Eds.)
Insights from Research in Science Teaching and Learning
Springer International Publishing, 2015, pp. 3-8
Book Series: Contributions from Science Education Research, Vol. 2, 2016, Series Editor: Robin Millar
ISBN: 978-3-319-20073-6 (Print) 978-3-319-20074-3 (Online)
DOI: 10.1007/978-3-319-20074-3
15. N. Papadouris and C. P. Constantinou
Distinctive features and underlying rationale of a philosophically-informed approach for energy teaching.
In R. F. Chen, A. Eisenkraft, D. Fortus, J. Krajcik, K. Neumann, J. Nordine and A. Scheff (editors) *Teaching and Learning of Energy in K-12 Education.* Springer Publications, 2014, pp. 207-222, ISBN: 978-3-319-05016-4, DOI 10.1007/978-3-319-05017-1

16. R. Pintó, M. Hernández and C. P. Constantinou
On the transfer of teaching-learning materials from one educational setting to another. In *C. Bruguiere, A. Tiberghien, P. Clement (editors) Topics and Trends in Current Science Education,* Contributions from Science Education Research Series. Springer Publications, 2014, pp. 535-552, ISBN: 978-94-007-7280-9.

IV. Publications in Conference Proceedings with Review System

17. Chr. Th. Nicolaou, B. Bredeweg, C. P. Constantinou and J. Lochem
Assessment of student-constructed models on topics relevant to heat and temperature systems.
In: C. Fazio and R. M. Sperandeo Mineo (Editors): Teaching/Learning Physics: Integrating Research into Practice.
Proceedings of the GIREP – MPTL 2014 International Conference, 7-12 July 2014, Palermo, Italy.
Publisher: Universita degli Study di Palermo, 2015, pp. 667-674.
ISBN: 978-88-907460-7-9.
18. Th. Petrou, Chr. Th. Nicolaou, P. Karnaou and C. P. Constantinou
Cognitive processes enacted by learners during co-construction of scientific models.
In Futschek, G. And Kynigos, C. (editors): Proceedings of the 3rd International Constructionism Conference Vienna, Austria, 19-23 August 2014, pp. 207-216, Oesterreichische Computer Gesellschaft. ISBN: 978-3-85403-301-1.

V. Other Publications

19. X. Danos, E. De Corte, Chr. Mappouridou, C. P. Constantinou,
30 years of EARLI: Personal Stories,
European Association for Research on Learning and Instruction, Leuven, Belgium, August 2015.

In: Teaching and Learning
Editor: Malcolm Vargas

ISBN: 978-1-63485-228-9

Chapter 5

STUDENTS' MISCONCEPTIONS IN LEARNING ALGEBRA: A DIAGNOSTIC TEACHING APPROACH

Wilfred W. F. Lau*
Department of Curriculum and Instruction,
The Chinese University of Hong Kong,
Shatin, NT, Hong Kong SAR, China

ABSTRACT

Students commonly hold misconceptions in learning algebra. Because algebra is generally regarded as a gatekeeper for advanced mathematical studies in both secondary school and university and it provides essential training for engineers, scientists, and mathematicians, identifying and rectifying these misconceptions is crucial during the learning process. This chapter describes and defines students' misconceptions in learning mathematics in general and discusses the behaviorist and constructivist views on learning misconceptions. Many errors and misconceptions of students related to algebraic manipulations were found to result from overgeneralization and overspecialization. This chapter specifically elucidates student algebraic errors and misconceptions in variables, expressions, equations, and word problems. It further illustrates through examples that diagnostic teaching, which

* Corresponding author: Tel: 852 39436941; fax: 852 26036724. E-mail: wwflau@cuhk.edu.hk.

greatly emphasizes students' misconceptions and the role of cognitive conflict in the teaching and learning process, can be used as an effective teaching intervention to promote conceptual changes in learning algebra.

Keywords: misconceptions, learning algebra, conceptual change, diagnostic teaching

HOW DO PEOPLE LEARN?

Learning occurs when learners' knowledge changes through experiences (Mayer, 2001). In general, learning involves strengthening correct responses; adding new information to the memory; and making sense of the presented material by attending to relevant information, followed by mentally reorganizing and connecting it with what is already known (Clark and Mayer, 2008, p. 33). Human learning can be explained through three major approaches: behavioral (learning occurs when a correct response is obtained after a specific environmental stimulus is presented), cognitive (learning is a change of the knowledge state and knowledge acquisition is an internal mental process, which involves remembering, retrieving, and storing information in memory by the learners), and constructivist (learning occurs when learners construct knowledge through their active interactions with the context). Cognition concerns how people perceive, process, store, and retrieve information. Thus, constructivism can be an appropriate and useful avenue for studying student cognition.

According to Piaget (1970), two major principles are involved in guiding the knowledge construction process: adaptation and organization. During adaptation, students attempt to respond to external stimuli (assimilation) and make continual effort to fit them (accommodation) into existing mental structures (schemas). Equilibration occurs when assimilation and accommodation are balanced. As new information arrives, existing schemas are organized and new schemas are developed for constructing new knowledge.

WHAT ARE MISCONCEPTIONS?

Confrey (1990) reviewed studies on misconceptions in three fields—science, mathematics, and programming—and reported various terms used to describe misconceptions in these fields: alternative conceptions, student conceptions, preconceptions, conceptual primitives, private concepts, alternative frameworks, systematic errors, critical barriers to learning, and naïve theories.

Leinhardt, Zaslavsky, and Stein (1990, p. 5) described and defined student misconceptions in learning mathematics in detail as follows:

> "Misconceptions are features of a student's knowledge about a specific piece of mathematics knowledge that may or may not have been instructed. A misconception may develop as a result of overgeneralizing an essentially correct conception, or may be due to interference from everyday knowledge. To qualify, a misconception must have a reasonably well-formulated system of ideas, not simply a justification for an error. So although misconception does not need to be an entire theory, it should be repeatable and/or explicit rather than random and tacit."

Graeber and Johnson (1991, pp. 3-15) further noted three characteristics of misconceptions: self-evident (there is no need to prove them), coercive (it is compelling to use them as initial responses), and widespread (they are commonly found among both naïve learners and more academically able students). Kaur (2013) identified numerous causes of misconceptions including lack of knowledge of concepts, erroneous teaching by teachers and textbooks, invalid transfer of daily experiences, confusion of terms, overgeneralization of concepts, misclassification and misidentification of concepts, and imprecise use of language.

VIEWS ON MISCONCEPTIONS

The behaviorist approach to learning regards student errors as problems that should be avoided and misconceptions as mistaken beliefs that should be replaced with accurate information (Olivier, 1996). A learner passively receives information from external stimuli and his or her behavior is influenced by positive or negative reinforcement. This approach assumes that knowledge can be imparted to a learner by another person and that the current knowledge possessed by the learner has no role in the learning process.

The constructivist approach to learning views knowledge as created through the interactions between cognitive structures called schemas and experiences (Smith, diSessa and Roschelle, 1993). Learning requires assimilation of information into existing schemas. Misconceptions occur when the information in existing schemas interacts with the newly assimilated information. Thus, misconceptions are inevitable, but they are valuable resources that can aid teachers in understanding what students know and how they construct knowledge. This approach recognizes the role of prior knowledge in the learning process.

HISTORICAL DEVELOPMENT OF ALGEBRA

Usiskin (1988) identified four fundamental conceptions of algebra: the first regards algebra as generalized arithmetic, the second views algebra as a study of procedures for solving some types of problems (e.g., word problems), the third considers algebra a study of relationships among quantities, and the fourth considers algebra a study of structures (e.g., abstract algebra). These four conceptions construe the notion of algebra in increasing sophistication.

Katz and Barton (2007) categorized the historical development of algebra in three stages: rhetorical, syncopated, and symbolic. At the rhetorical stage, ordinary language was used to express problems and their solutions; for example, the rhetorical form "The thing plus two equals five" represented the equation "$x + 2 = 5$" in modern algebra. At the syncopated stage, some symbols were used to denote unknown but specific quantities in a restrictive sense. Diophantus, who is often deemed the father of algebra, started using symbols in writing equations; for example, "$n + 5 = 13$" with n representing a specific unknown. At the symbolic stage, full symbolism was developed, and letters were used to represent both unknown and given quantities as parameters. This provides general solutions to problems and enables generalization of patterns; for example, the general form of the equation of a line can be written as "$ax + by + c = 0$" with a, b, and c being parameters and x and y being variables. Katz and Barton (2007, p. 186) also identified four conceptual stages with respect to changes in algebraic expressions:

> "the geometric stage, where most of the concepts of algebra are geometric; the static equation-solving stage, where the goal is to find numbers satisfying certain relationship; the dynamic function stage, where motion seems to be an underlying idea; and finally the abstract stage, where structure is the goal."

Algebra is generally regarded as a gatekeeper for advanced mathematical study in both secondary school and university. It provides essential training to people engaging in careers such as engineering, science, and mathematics.

ALGEBRAIC ERRORS AND MISCONCEPTIONS

Although errors and misconceptions can be different, misconceptions generally lead to errors. In algebraic manipulations, most errors and misconceptions of students frequently result from overgeneralization and overspecialization (Ashlock, 2010). Regarding overgeneralization, some students believe that a sum is the number written on the right side of an equality sign, regardless of whether the operation is addition or subtraction. Furthermore, some students consider $2y$ to be $20 + y$ possibly because they overgeneralize from expressions such as $23 = 20 + 3$. Other examples of overgeneralization from the distributive law include $(a + b)^3 = a^3 + b^3$, $\sin(a + b) = \sin a + \sin b$, and $\log(a + b) = \log a + \log b$. Regarding overspecialization, some students believe that to add two decimals, the number of digits on either side of the decimal point must be the same, and thus, they write 100.36 + 12.57 as 100.36 + 125.70. Some students believe that multiplication and division of fractions require like denominators as addition and subtraction of fractions do.

This chapter discusses student algebraic errors and misconceptions in variables, expressions, equations, and word problems. Macgregor and Stacey (1997) revealed evidence of student misinterpretations (Years 7–10) of algebraic letters as either generalized numbers or specific unknowns. These misinterpretations originated from "intuitive assumptions and pragmatic reasoning about a new notation, analogies with familiar symbol systems, interference from new learning in mathematics, and the effects of misleading teaching materials" (p. 1). For example, some students wrote that h10 indicated "add 10 to h" and 1y indicated "take 1 from y", assuming that their knowledge from the Roman numeration system in which VI means "1 more than 5" and IV means "1 less than 5" is valid (analogies with familiar symbol systems). Some students used their knowledge regarding right-angled triangles and Pythagoras's theorem and wrote $x^2 + 5^2 + 8$ as the distance around a pentagon with sides x cm, x cm, 5 cm, 5 cm, and 8 cm (interference from new learning in mathematics).

Conjoining letters in algebraic expressions is a commonly observed error among students. Some students are confused regarding the dual nature of mathematical notations as processes and objects (Sfard, 1991). For example, $5x + 3$ can be understood as adding 5 times x and 3 or as an object $5x + 3$. Some students may simplify $5x + 3$ to $8x$ to obtain a final answer without realizing that it is actually an object. Some students likely perceive an object as incomplete and therefore oversimplify it to obtain a final answer. Although this practice is reinforced in many arithmetic problems, it may cause inappropriate transfer to algebra learning.

The equality sign in an equation has two relational interpretations. In a symmetric relation, it shows equality of quantities on both sides of the sign. In a transitive relation, it signifies that a quantity on one side can be moved to the other side according to some algebraic rules. However, students often interpret the equality sign as an operational symbol, rather than a relational symbol, possibly because of the influence of their early arithmetic learning experiences (Powell, 2012). For example, some students may solve a problem such as 4 + 3 = ___ + 5 by adding all three numbers and obtain an answer of 12. Some students may answer 7 in the blank space for 4 + 3 = ___ + 3 because they believe that the answer should be after the equality sign.

When solving algebraic word problems, students face major difficulty in comprehending a problem and formulating an equation or set of equations (Stacey and MacGregor, 2000). This process involves assigning unknowns, identifying constants, representing relationships among unknowns and constants, and finding the unknowns. For students, translating a linguistic expression in a word problem into a symbolic expression in an equation is particularly difficult. Kieran, Booker, Filloy, Vergnaud, and Wheeler (1990) revealed that for solving a word problem, some students tend to focus on the operators used to solve the problem, rather than representing the problem situation mathematically. In this case, the previously learned arithmetical procedures are likely to interfere with the development of subsequent algebraic concepts.

Diagnosis of Misconceptions

Two methods are commonly used to diagnose misconceptions in mathematics, namely individual interviews and classification of errors on collective assessments related to particular misconceptions. During individual interviews, students explain and justify their answers when solving problems. Interviews with students can be effective in understanding students' mathematical concepts, skills, and dispositions. They enable teachers to gain insights into how students understand concepts and procedures, to pinpoint any misconceptions and errors, to examine how students reason, to know how students are able to communicate mathematical ideas well, and to understand students' attitudes toward mathematics (Ashlock, 2010, p. 25).

In their study, Vamvakoussi and Vosniadou (2004) asked students to think aloud and comment on their answers for questions such as "how many numbers are there between 0.005 and 0.006?" In another study, for the same question, Vamvakoussi and Vosniadou (2010) provided students with alternatives including "there is no other number," "there is a finite number of decimals," "there is a finite number of fractions," "there are infinitely many decimals," "there are infinitely many fractions," "there are infinitely many numbers and they can have various forms," and "I don't agree with any of the above. I believe that…" The researchers classified student responses on the multiple-choice items according to their particular misconceptions.

Conceptual Change and Mathematics Learning

Skemp (1987) identified two main categories of mathematical understanding: (a) instrumental understanding requires knowing what to do in a mathematical task, whereas (b) relational understanding requires both knowing what to do in a mathematical task and knowing why the task is performed in a particular manner. Hiebert and Carpenter (1992) defined two similar categories: (a) procedural understanding is instrumentally oriented and includes knowing how, but not why, and (b) conceptual understanding is relationally oriented and includes knowing both how and why. Conceptual understanding is considered more critical than procedural understanding in learning mathematics, and conceptual change is usually required to establish conceptual understanding.

Conceptual change refers to "the kind of learning required when the new information to be learned comes into conflict with learners' prior knowledge usually acquired on the basis of everyday experiences" (Vosniadou and Liven, 2004, p. 445). It is a learning process requiring major reorganization of existing knowledge to enable understanding of new knowledge. For example, students understand the concept of counting numbers and of operations, such as addition and subtraction, applied to those numbers during the early years. They also know that each positive integer is followed by a number. However, when they learn that fractions and decimals are numbers, they must restructure their concept of numbers from counting numbers to rational numbers and incorporate several changes, including developing a mathematical understanding of division (Tirosh and Tsamir, 2006).

Through their elementary school experiences, students learn that multiplication always results in a greater number. Later, when they realize that this principle does not apply to the cases of multipliers less than 1, "they will continue to suggest that $0.2 \times 0.3 = 0.6$ despite being taught methods which give the result 0.06" (Kennewell, 1994, p. 3), and this restructuring of the concept of multiplication of numbers is required.

DIAGNOSTIC TEACHING FOR CONCEPTUAL CHANGE

Diagnostic teaching, also known as conflict teaching, greatly emphasizes student misconceptions and the role of cognitive conflict in the teaching and learning process. The diagnostic teaching approach is based on "identifying the key conceptual points and common misconceptions within the topic, using existing research, and testing directly the students who are to be taught. The teaching is then designed to focus on these points, giving the students substantial challenges, provoking cognitive conflict by exposing misconceptions, and resolving them through discussion" (Bell, 1993, p. 115). The approach is characterized by the following principles: lessons focus on specific difficulties, questions create a tension or cognitive conflict that requires resolution, activities provide opportunities for meaningful feedback to the students, lessons include time for entire class discussion, and opportunities are provided for pupils to "consolidate" what has been learned (Bell, 2005, pp. 1-2).

Chow and Treagust (2013) designed a conflict teaching lesson to provide Form 2 (Grade 8) students with an introductory task aiming to confront students with a problem having a conceptual obstacle. The students were then asked to solve the problem. The authors created cognitive conflict by using a counter example or two conflicting examples. The teacher explained to the students an alternative conception through discussion of why one method does not work or why one method is superior to another. The authors illustrated their ideas with a "think of a letter" activity, which addressed the students' misconception on the order of operations that algebraic expressions must always be read from left to right.

A comparison between student achievement in a pretest and a posttest administered after 6 weeks showed the effectiveness of this teaching intervention. Results indicated a statistically significant improvement in student achievement in mathematics [for pretest, mean (M) = 13.06, standard deviation (SD) = 3.54; for posttest, M = 14.91, SD = 3.94]. The effect size was 0.49, indicating that students learning through this approach on average showed improved achievement by almost one half of a SD.

Fraser (2007) studied the effect of cognitive conflict interventions on Grade 11 students' structural understanding of algebra. All participants had a procedural understanding of algebra before the interventions. Questions for testing structural understanding included quadratic inequalities that have no solution or an infinite number of solutions, equations and inequalities or systems of equations and inequalities that have an infinite number of solutions, parametric equations, pairs of equations and inequalities that in some cases are equivalent and not transformable procedurally from one to the other, and pairs of equations and inequalities that are not equivalent but are tempting for students to transform from one to the other (pp. 37-38).

The cognitive conflict interventions were designed considering the following points: the students were at an induced conflict level, a problem was presented that induced the conflict level, the interventions occurred in a nonintimidating environment, and an alternate understandable conception was provided. The researcher moved around the classroom to identify evidence that cognitive conflict was triggered among the students, such as recognition of an anomalous result, interest in the result, anxiety, and reappraisal of the situations.

The students performed much higher in a questionnaire and interview after the intervention than the ones before. They demonstrated a structural understanding of algebra concepts as reflected in their language use to describe the problem-solving process, their ability to oscillate between procedural and structural views of a problem, their explicit justification of the method used to solve an unfamiliar problem, and use of a suitable control mechanism to ensure that a correct answer is obtained. Diagnostic teaching has been suggested to result in more effective long-term learning than standard "positive only" methods, which avoid the analysis and understanding of students' misconceptions (Askew and Wiliam, 1995).

Lee and Kwon (2001) reviewed different types of cognitive conflicts in the literature such as (a) internal cognitive conflict (between two competing ideas), (b) external social conflict (between two external events or sources of information), and (c) internal–external conflict (between internal and external events) (p. 5). Because the types of cognitive conflict vary and they may require different levels of conflict for an appropriate solution, cognitive conflict can have constructive, destructive, or meaningless potential (Fraser,

2007). This implies that the use of diagnostic teaching in the classroom should be planned far in advance to produce the appropriate conceptual change.

CONCLUSION

Students commonly have misconceptions in learning algebra. However, some teachers, mainly those with behaviorist views of learning, believe that student errors should be avoided and replaced with correct information, whereas others may consider misconceptions to be meaningful teaching resources for teachers to understand how students construct or misconstruct knowledge from a constructivist learning perspective. In this chapter, I illustrated through examples that diagnostic teaching can be used as an effective teaching intervention to promote students' conceptual change in learning algebra. To benefit from this promising intervention, I strongly advise teachers to identify students at risk of having misconceptions regarding algebra and introduce the intervention at the early stage.

REFERENCES

Ashlock, R. B. (2010). *Error patterns in computation: Using error patterns to help each student learn* (10th edition). Boston : Allyn and Bacon.

Askew, M. and D. Wiliam (1995). Recent Research in Mathematics Education 5-16. London, HMSO.

Bell, A. (1993). Some experiments in diagnostic teaching. *Educational Studies in Mathematics*, 24(1), 115 – 137.

Bell, A. (2005). Introduce diagnostic teaching. Alan Bell and the Toolkit Team. *A Strategy in the Toolkit for Change Agents*. MARS, Michigan State University.

Chow, T.-C. and Treagust, D. (2013). An intervention study using cognitive conflict to foster conceptual change. *Journal of Science and Mathematics Education in Southeast Asia*, 36(1), 44-64.

Clark, R. C. and Mayer, R. E. (2008). *E-Learning and the science of instruction: Proven guidelines for consumer and designer of multimedia learning*. (2nd ed.) San Francisco, CA: Pfeiffer.

Confrey, J. (1990). A review of the research on student conceptions in mathematics, science and programming. In C.B. Cazdan (Ed.), *Review of*

Research in Education, (Vol. 16) (pp. 3–56). Washington, DC: American Educational Research Association.

Fraser, D. (2007). Using cognitive conflict to promote structural understanding of Grade 11 algebra. (Master of Science Thesis), Simon Fraser University, Canada. Retrieved from http://www.peterliljedahl.com/wp-content/uploads/Thesis-Duncan-Fraser.pdf.

Graeber, A. and Johnson, M. (1991). *Insights into secondary school students' understanding of mathematics.* College Park, University of Maryland, MD.

Hiebert, J. and Carpenter, T. (1992). Learning and teaching with understanding. In D. Grouws. (Ed.), *Handbook of Research on Mathematics Teaching and Learning.* (pp. 65-97). NY: Macmillan Publishing Company.

Katz, V. J. and Barton, B. (2007). Stages in the history of algebra with implications for teaching. *Educational Studies in Mathematics*, 66(2), 185-201.

Kaur, G. (2013). A review of selected literature on causative agents and identification strategies of students' misconceptions. *Educationia Confab,* 2(11), 79-94.

Kennewell, S. (1994). A computer's role in revealing and influencing learners' understanding. *Paper presented at the Annual Conference of the Australian Association for Research in Education*, Newcastle, Australia.

Kieran, C., Booker, G., Filloy, E., Vergnaud, G. and Wheeler, D. (1990). Cognitive processes involved in learning school algebra. In P. Nesher and J. Kilpatrick (Eds.), *Mathematics and cognition* (pp. 96-112). Cambridge University Press.

Lee, G. and Kwon, J. (2001). What do you know about students' cognitive conflict: A theoretical model of cognitive conflict process? Proceedings of 2001 AETS Annual Meeting, Costa Mesa, CA, (pp. 1 -19). (ERIC Document Reproduction Service No. ED472903).

Leinhardt, G., Zaslavsky, O. and Stein, M. K. (1990). Functions, graphs, and graphing: Tasks, learning and teaching. *Review of Educational Research,* 60(1), 1-64.

Macgregor, M. and Stacey, K. (1997). Students' understanding of algebraic notation: 11-15. *Educational Studies in Mathematics*, 33, 1–19.

Mayer, R. E. (Ed.). (2001). *Multimedia learning*. New York: Cambridge University Press.

Olivier, A. (1996). Handling pupils' misconceptions. *Pythagoras,* 21, 10–19.

Piaget, J. (1970). *Genetic epistemology.* New York: Columbia University Press.

Powell, S. R. (2012). Equations and the equal sign in elementary mathematics textbooks. *Elementary School Journal,* 112(4), 627–648.

Sfard, A. (1991). On the dual nature of mathematical conceptions: Reflections on processes and objects as different sides of the same coin. *Educational Studies in Mathematics*, 22, 1–36.

Skemp, R. R. (1987). *The psychology of learning mathematics.* USA: Erlbaum Associates.

Smith, J., Disessa, A. and Roschelle, J. (1993). Misconceptions reconceived: A constructivist analysis of knowledge in transition. *The Journal of the Learning Sciences,* 3(2), 115-163.

Stacey, K. and MacGregor, M. (2000). Learning the algebraic method of solving problems. *Journal of Mathematical Behaviour*, 18(2), 149–167.

Tirosh, D. and Tsamir, P. (2006). Conceptual Change in mathematics learning: The case of infinite sets. In J. Novotna, H. Moraova, M. Kratka and N. Stehlikova (Eds.), Proceedings of the 30th Conference of the International Group for the Psychology of Mathematics Education (Vol. 1) (pp. 159-161). Prague, Czech Republic: PME.

Usiskin, Z. (1988). Conceptions of school algebra and uses of variables. In A. F. Coxford and A. P. Shulte (Eds.), *The ideas of algebra*, K – 12: 1988 yearbook (pp. 8-19). National Council of Teachers of Mathematics.

Vamvakoussi, X. and Vosniadou, S. (2004). Understanding the structure of the set of rational numbers: A conceptual change approach. *Learning and Instruction*, 14(5), 453-467.

Vamvakoussi, X. and Vosniadou, S. (2010). How many decimals are there between two fractions? Aspects of secondary school students' understanding of rational numbers and their notation. *Cognition and Instruction,* 28(2), 181-209.

Vosniadou, S. and Lieven, V. (2004). Extending the conceptual change approach to mathematics learning and teaching. *Learning and Instruction,* 14(5), 445-451.

BIOGRAPHICAL SKETCH

Wilfred W.F. Lau

Affiliation: Department of Curriculum and Instruction, The Chinese University of Hong Kong, Shatin, NT, Hong Kong SAR, China

Education: PhD

Address: Department of Curriculum and Instruction, The Chinese University of Hong Kong, Shatin, NT, Hong Kong SAR, China

Research and Professional Experience: Post-doctoral Fellow

Professional Appointments: Assistant Professor

Publications Last 3 Years:

Yuen, A. H. K., Lau, W. W. F., Park, J. H., Lau, G. K. K. and Chan, A. K. M. (2016). Digital equity and students' home computing: A Hong Kong study. *The Asia-Pacific Education Researcher*. (SSCI; IF: 0.793).

Hui, C. H., Lau, W. W. F., Cheung, S. H., Cheung, S.-F., Lau, E. Y. Y. and Lam, J. (2015). Predictors and outcomes of experiences deemed religious: A longitudinal investigation. *International Journal for the Psychology of Religion, 25*(2), 107-129. (SSCI; IF: 1.095).

Hui, C. H., Lau, E. Y. Y., Lam, J., Cheung, S. F. and Lau, W. W. F. (2015). Psychological predictors in Chinese Christians' church attendance and religious steadfastness: A three-wave prospective study. *Psychology of Religion and Spirituality, 7*(3), 250-264. (SSCI; IF: 1.500).

Lau, W. W. F., Hui, C. H., Lam, J., Lau, E. Y. Y., Ng, D. and Cheung, S. F. (2015). Psychometric evaluation of the Spiritual Transcendence Scale in a Chinese sample: Is there factorial invariance across gender, occupation, and religion? *International Journal for the Psychology of Religion*. (SSCI; IF: 1.095).

Lau, W. W. F., Hui, C. H., Lam, J., Lau, E. Y. Y. and Cheung, S.-F. (2015). The relationship between spirituality and quality of life among university students: An autoregressive cross-lagged panel analysis. *Higher Education, 69*(6), 977-990. (SSCI; IF: 1.124).

Lau, W. W. F. and Yuen, A. H. K. (2015). Factorial invariance across gender of a perceived ICT literacy scale. *Learning and Individual Differences, 41*, 79-85. (SSCI; IF: 1.565).

Chu, S. K. W., Lau, W. W. F., Chu, D. S. C., Lee, C. W. Y. and Chan, L. L. H. (2014). Media awareness among Hong Kong primary students. *Journal of Librarianship and Information Science.* (SSCI; IF: 0.273).

Lau, W. W. F. and Yuen, A. H. K. (2014). Developing and validating of a perceived ICT literacy scale for junior secondary school students: Pedagogical and educational contributions. *Computers and Education*, 78, 1-9. (SSCI; IF: 2.630).

Lau, W. W. F. and Yuen, A. H. K. (2014). Internet ethics of adolescents: Understanding demographic differences. *Computers and Education*, 72, 378-385. (SSCI; IF: 2.630).

In: Teaching and Learning
Editor: Malcolm Vargas
ISBN: 978-1-63485-228-9

Chapter 6

EFFECTS OF LEVEL OF REGULATORY TEACHING ON ACHIEVEMENT EMOTIONS IN THE LEARNING PROCESS: ANXIETY AND COPING STRATEGIES IN HIGHER EDUCATION

Jesús de la Fuente[1,*], Paul Sander[2,≠], María Cardelle-Ellawar[3,#] and Silvia Pignata[4,±]

[1]Developmental and Educational Psychology, University of Almería, Almería, Spain
[2]Department of Psychology, Arden University, Coventry, UK
[3]Educational Psychology, Arizona State University, Tempe, Arizona, US
[4]School of Engineering, University of South Australia, Australia

ABSTRACT

Introduction. In the teaching-learning process at university, achievement emotions prove to be very important and are attracting much attention as an important student variable. *Regulatory Teaching* is a

---

* Full Professor of Developmental and Educational Psychology
≠ Associate Professor
Emeritus Professor of Educational Psychology
± Doctor in Work Psychology

characteristic of effective teaching that is related to students' achievement emotions. *Test anxiety* is a type of negative emotion, present in study and exam situations. *Coping strategies* is a meta-emotional variable defined as behavioral and cognitive efforts that are carried out in order to deal with stress, whereby we can address both external and internal stress-generating demands. The *objective* of the present research is to establish how different levels of Regulatory Teaching produce differences in test anxiety and the types of coping strategies used by university students.

Method. A total of 202 students from the University of Almeria (Spain) participated in an assessment of the teaching-learning process and achievement emotions. The following measures were used: (1) *Regulatory teaching* with the IATLP, in its Spanish validated version; (2) *Test anxiety*, with the TAI-80, in the Spanish validated version, and (3) *Coping strategies*, with the Spanish Coping Strategies Scales, also in a validated version. The analyses undertaken were cluster analyses (low-medium-high), ANOVAs and MANOVAs, with an ex post-facto design.

Results. The level of Regulatory Teaching had an effect on level of test anxiety, and coping strategies. In this study, the university students who perceived a *high level of Regulatory* Teaching had a low level of test anxiety, a low level of emotion-focused coping strategies, and a high level of problem-focused strategies. However, students who perceived a *low level of Regulatory* Teaching had a high level of test anxiety, high level of emotion-focused coping strategies, and a low level of problem-focused strategies.

Conclusion. Evidence is offered to defend that *Regulatory Teaching* is an important characteristic of effective teaching, because: (1) it protects against strong negative emotional reactions; (2) it is associated with *perceived predictability* in the academic context; (3) consequently, the student focuses on the problem, and does not need to focus on negative achievement emotions. With this result, we can focus intervention strategies on the design and development of regulatory teaching, in the university academic context.

Keywords: regulatory teaching, effective teaching, achievement emotions, text anxiety, coping strategies

INTRODUCTION

In the teaching-learning process at university, *achievement emotions* prove to be very important and are attracting much attention as an important student variable. Pekrun and Stephen (2012) have established future directions for research on achievement emotions. During the last ten years, studies have been carried out to establish the nature and type of emotions. Research on interventions is also important; there is plentiful evidence focusing on the study of anxiety, but not addressing the other negative emotions (anger, hopelessness, etc.), for the purpose of replacing them with positive emotions (hope, pride, engagement in learning). Research suggests that it is possible to design effective learning environments that relieve anxiety. The design of such learning environments for positive emotionality remains to be established.

Regulatory Teaching

Theoretical contributions and empirical evidence can help psychological and educational practice through learning environments that offer channels for preventing and treating student anxiety.

Regulatory teaching (de la Fuente, 2011; de la Fuente and Justicia, 2007; de la Fuente et al., 2014, 2015) refers to encouragement of self-regulation in students, and is characteristic of effective teaching. Research has taken many approaches to *effective teaching* (Goe, Bell and Little, 2008, for a review). In empirical research, effective, high quality teachers are those who have a positive impact on their students' engagement with learning activities (Reeve, Jang, Carrell, Jeon and Barch, 2004), as well as on students' performance in relation to learning (self-regulation, social competencies, academic achievement). It is very important to consider factors that mediate in students' performance (Roehrig and Chistesen, 2010):

1) *Organization of the content and activities*. Keeping students' attention and interest is not only a sign of good teaching, but also of good planning and management of one's teaching. Several strategies can be used for this purpose. In addition, good teaching involves a good selection of the content and knowledge to be learned (Roeghrig and Chistesen, 2010). The material and curricular activities may also contribute to stimulating interest in the content. When students have interest in a new activity it is because they were previously successful in similar situations in the past (Guthrie, McRae and Klauda, 2007). Teachers can help make this happen. Also, students show greater engagement in practical activities where they have to solve applied problems (Roeghrig and Chistesen, 2010). Therefore, these strategies

can help teachers to improve their classroom management, encouraging their students' motivational and cognitive production.

2) *Planning for the majority of the class.* One of the components of planning is good, flexible organization, as a part of student engagement. An effective teacher is a good organizer, anticipating problems and seeking planning alternatives (Roehrig and Chistersen, 2010). Also, he or she plans for success, using a variety of instructional strategies in each lesson. When comparing new teachers with experts, the latter differ in terms of greater complexity in elaborating the learning content, in how they respond automatically to planning-related situations, in the decision-making process (forming student groups, selecting work material, the innumerable decisions that are made in adapting to the class) (Corno, 2008) and in scheduling the tasks. When teachers plan well and use good methods, students are more engaged (Roehrig and Chistesen, 2010).

3) *Encouraging deep processing and self-regulation.* Self-regulation is the ability to control one's own behavior, impulses, emotions and thoughts (including attentional processes), thus leading to a deep learning approach (critical, interactive), as opposed to a surface approach (passive, memorization-based) (Entwistle and Entwistle, 1991). Fundamentally, this practice refers to promoting metacognition in students, and the knowledge of how to monitor their own cognitive processes. Metacognition is the highest order of thought, and can be implemented in students' learning through teaching. The highest level of cognitive engagement (the teaching of thought) is a good level of emotional engagement (positive atmosphere), in interaction with other behavioral engagement behaviors (class management), as a means of getting others engaged (Fredricks, Blumenfeld and Paris, 2004). Teacher behaviors that predict self-regulated learning are modeling help, and activities in the zone of proximal development. These practices for helping students demonstrate the content and skills needed for applying self-regulation to thoughts, behaviors and affect (Zimmerman, 1989).

Test Anxiety

There is general consensus in understanding emotions to be a multidimensional phenomenon. Emotions include affective, cognitive, physiological, motivational and expressive processes. Emotions can be grouped according to a three-fold taxonomy: 1) according to the object of their focus (activity-focus); 2) their valence (positive-negative); and 3) degree of activation (activation-disactivation) (Pekrun, 1992, 2006, 2009). Achievement emotions may refer to the execution of the activity or to its conclusion, experiencing either success or failure. Assessment of achievement emotions involves the use of different instruments, whether self-reports, neurological, physiological, experimental or observational, both verbal and nonverbal. Traditionally, such assessment has focused on test anxiety. This measurement, however, despite its interest, is limited in terms of the range of possible emotions, positive and negative. There is therefore a need to move on to broader instruments that include assessment through self-reporting a variety of emotions: enjoyment, hope, pride, relief, anger, anxiety, shame, hopelessness and boredom. These emotions are assessed in different situations: listening in class, studying, preparing for the test, doing the test. Emotions assessed by the AEQ predict students' academic performance, as well as their interest, goals, learning strategies used, effort in study, and academic self-regulated learning (Acee et al., 2010).

Coping Strategies

A *meta-emotional variable* (de la Fuente, 2015; De la Fuente, García-Torrecillas and Rodríguez-Vargas, 2015) is defined as the behavioural and cognitive efforts that are carried out in order to deal with stress, whereby we can address both external and internal stress-generating demands. From the point of view of the effects of stress, Lazarus and Folkman (1984) defined *coping strategies* as constant changes in cognitive processes and behavior, undertaken in order to manage specific external and/or internal demands that one considers to exceed or exhaust perceived resources. These responses make it possible to manage and somehow reduce the negative qualities of a stressful situation, through an effort to manage stress factors. These strategies may be considered meta-motivational variables for coping with stress. Keeping this in mind, they have proposed two coping styles: 1) emotion-focused, trying to reduce tension, physical activation and emotional reaction, and 2) problem-focused, trying to modify the problem situation in order to make it less stressful.

Aims

The *objectives* of the present research is to establish how different levels of Regulatory Teaching produce differences in the levels of test anxiety and types of coping strategies used by university students. According to previous evidence supplied by the DEDEPRO Model (de la Fuente, 2011; de la Fuente and Justicia, 2007) and the premises of the Theory of Self- vs. Other-Regulated Learning® (de la Fuente, 2014), we hypothesize that: (1) the level of regulatory teaching (low-medium-high) would be associated with, in interdependent fashion, the students' level of anxiety; hence, a lower level of regulatory teaching is expected to bring with it a higher level of test anxiety (worry and emotionality); (2) in complementary fashion, the level of regulatory teaching would correspond to the level and type of coping strategies used; a higher level of regulatory teaching is expected to seen in conjunction with students using a greater number of problem-focused strategies, and fewer emotion-focused strategies.

METHOD

Participants

A total of 202 students from the University of Almeria (Spain) participated in an assessment of the teaching-learning process and achievement emotions. The participants were in the second or fourth years of a Psychology degree program: 86.5% were women and 13.5% were men. Their ages ranged from 19 to 49, with a mean age of 23.08 ($\sigma_X = 4.4$).

Instruments

Regulatory teaching. The *Interactive Assessment of the Teaching and Learning Process*, *IATLP-D1* (De la Fuente and Martínez-Vicente, 2007), evaluated the teaching-learning process from the students' perspective and possible causal relations among the presage-process-product variables. Overall reliability for this scale was alpha = .75, which we considered acceptable. The IATLP-D1 (regulatory teaching) is a self-report instrument to be completed by the teacher, available in Spanish and English versions. Responses are on a Likert-type scale, with scores ranging from 1 (totally disagree) to 5 (totally agree). As for the instrument's external validity, results are also consistent, as there are different interdependent relationships among the perception of variables that exist in an academic environment. Variations in scores for the learning process have shown that the Scale is sensitive to the latter's influence on the teaching-learning process. These results provide empirical evidence of the scale's external validity (de la Fuente et al., 2012).

Test anxiety. The TAI-80, *Test Anxiety Inventory*, was used. This questionnaire is a reduced, validated Spanish adaptation of the STAI (State Trait Anxiety Inventory), by Spielberger (1980). This inventory provides a measurement of anxiety in test situations, addressing the two components explained above. The test contains two parts, with 10 questions each. Worry is evaluated through the existence of interfering and automatic thoughts that prevent the proper functioning of attention, working memory and performance in the assessment situation. Emotionality is evaluated through negative emotions, which can also be interfering. Reliability statistics are Alpha = .919, Guttman Split Half =.865 for worry, and Alpha = .819, Guttman Split Half = .834 for emotionality (Spielberger, 1980).

Coping strategies. The *Escala Estrategias de Coping* [Coping Strategies Scale], EEC, was used in its original version (Chorot and Sandín, 1987), as validated for university students (De la Fuente, 2014). Although the original instrument contained 90 items, the validation produced a first-order structure of 64 items and a second order with 10 factors and two dimensions, both of them significant, with adequate fit values in the latter case (Chi-square = 878.750; degrees of freedom (77-34) = 43, $p < .001$; NFI = .901; RFI = .945; IFI = .903, TLI = .951; CFI = .903). Reliability measures are Cronbach alpha of .93 (complete scale), .93 (first half) and .90 (second half), Spearman-Brown of .84 and Guttman of .80.

Procedure

The three instruments were completed voluntarily in a classroom situation, at the beginning, middle and end of the course, respectively. The online utility, *e-Coping with Academic Stress*, was used (De la Fuente, 2015; De la Fuente et al., 2014).

Data Analysis

The analyses employed were cluster analyses (low-medium-high) in the regulatory teaching variable, ANOVAs and MANOVAs, with an ex post-facto design.

RESULTS

Regulatory Teaching and Anxiety

A significant main effect appeared for the variable level of regulatory teaching (low-medium-high) on students' level of total test anxiety [$F(2,203) = 3.359$, $p < .05$, $eta^2 = .03$ ($3 < 1$, $p < .05$)]. The same was true of its components, [$F(4,402) = 2.922$, $p < .05$, $eta^2 = .028$], that is, for both the worry factor [$F(2,203) = 4.452$, $p < .01$, $eta^2 = .047$ ($3 < 1$, $p < .01$; $3 < 2$, $p < .05$)] and the emotionality factor [$F(2,203) = 2.918$, $p < .05$, eta2 = .02 ($3 < 1$, $p < .05$)], although the effect on the first factor is seen more clearly (Figure 1).

Regulatory Teaching and Coping Strategies

Regulatory Teaching and Coping Dimensions

A significant main effect appeared for the variable level of regulatory teaching (low-medium-high) on the dimensions of coping strategies [$F(4,332) = 2.779$, $p < .05$, $eta^2 = .032$], although there was a significant partial effect for problem-focused strategies [$F(2,168) = 3.609$, $p < .05$, $eta^2 = .042$; $3 > 1$, $p < .05$], and none for emotion-focused strategies [$F(2,618) = 0.838$, n.s.] (Figure 2).

Regulatory Teaching and Coping Strategies Factors

However, the level of the regulatory teaching variable (low-medium-high), showed a significant main effect on the factors of the coping strategies [$F(26, 308) = 1.939$, $p < .005$, $eta^2 = .141$]. Hence, with higher levels of regulatory teaching, there is greater use of problem-focused strategies and less use of emotion-focused strategies.

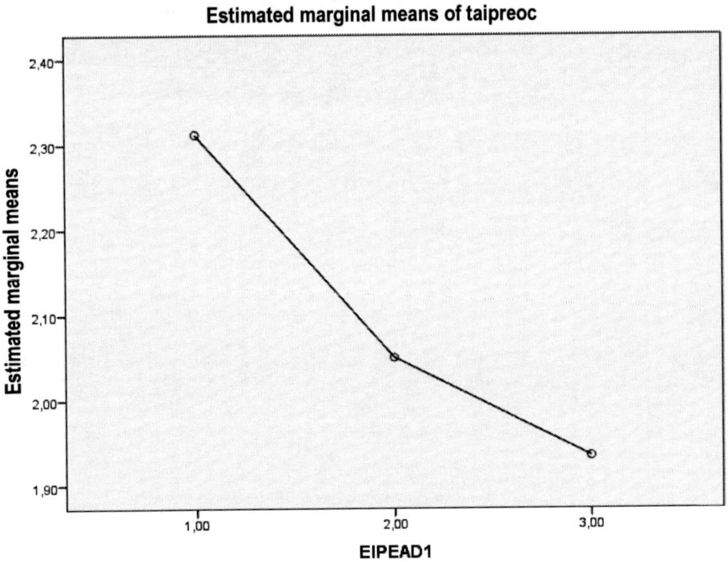

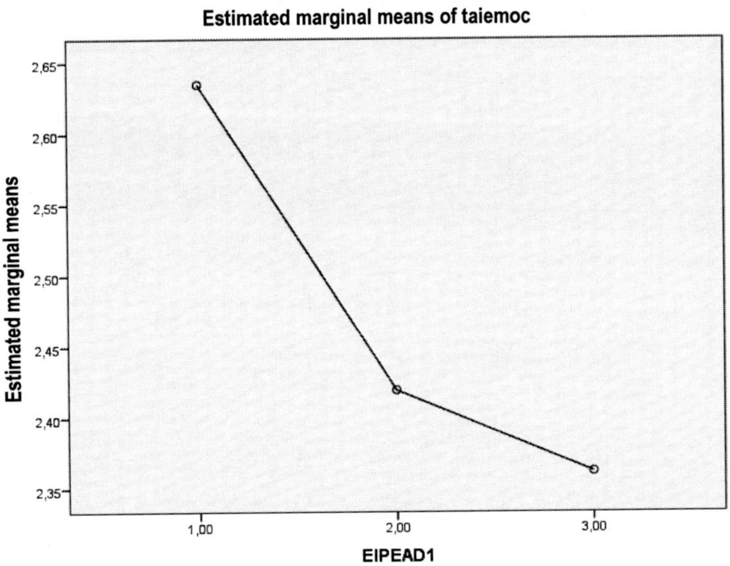

Note: (1) Low level, (2) Medium level; (3) High level; taipreoc = Worry; taiemoc = Emotionality.

Figure 1. Estimated Marginal Means. Levels of the components of test anxiety (worry, on the left, and emotionality, on the right), according to the level of regulatory teaching (EIPEAD1).

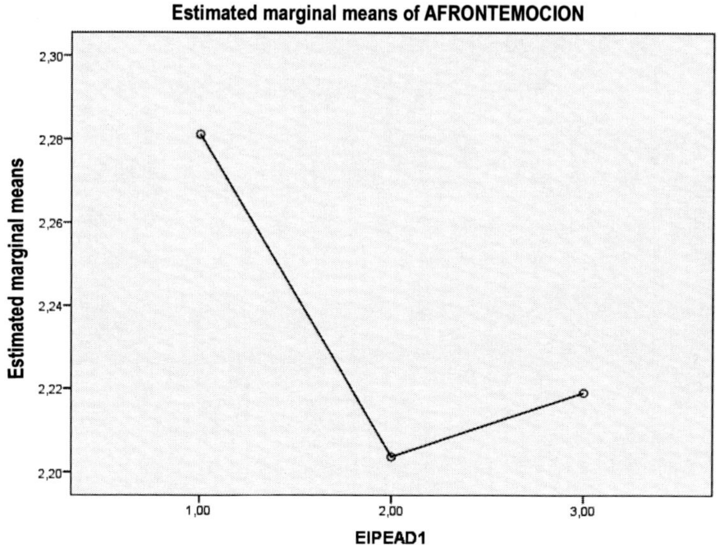

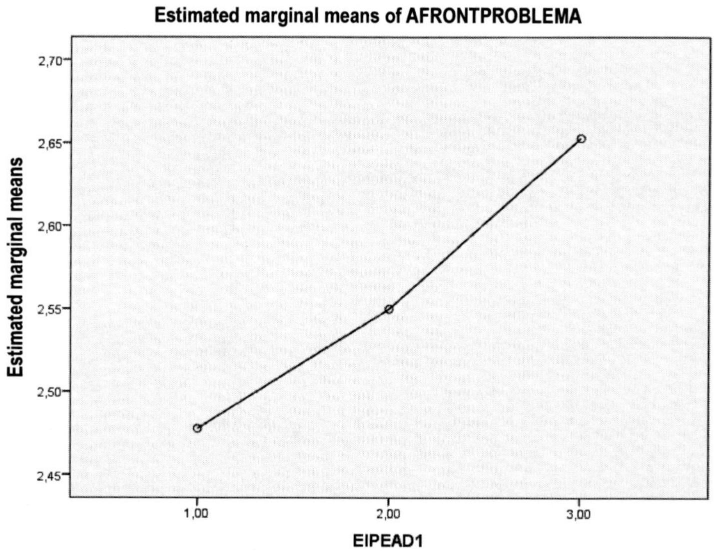

Note: (1) Low level; (2) Medium level; (3) High level; AFRONTEMOCION= emotion-focused coping; AFRONTPROBLEMA= problem-focused coping.

Figure 2. Estimated marginal means. Levels of the coping strategy dimensions (emotion-focused coping, on the left, and problem-focused coping strategies, on the right), according to the level of regulatory teaching (EIPEAD1).

1) *Problem-focused strategies*

The level of regulatory teaching affected the level of *problem-focused* coping strategies: *EEF2* [$F(2,165) = 5.280$, $p < .005$, $\text{eta}^2 = .060$; $3 < 1$, $p < .01$], *EECF5* [$F(2,165) = 1.154$, $p < .001$, $\text{eta}^2 = .086$, $3 < 1$, $p < .001$], *EEF6* [$F(2,165) = 3.015$, $p < .05$, $\text{eta}^2 = .035$, $3 < 2$, $p < .05$], *EEF10* [$F(2,165) = 4.483$, $p < .01$, $\text{eta}^2 = .052$, $3 < 2$, $p < .01$] and *EECF12* [$F(2,165) = 5.055$, $p < .01$, $\text{eta}^2 = .058$, $3 < 1$, $p < .001$]) (Figure 3).

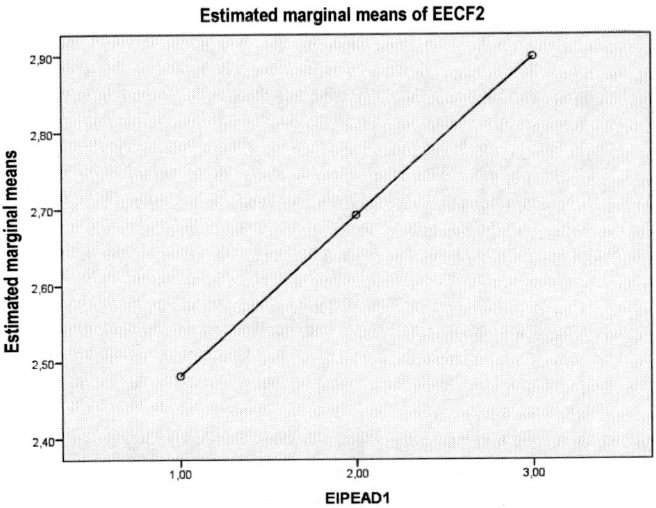

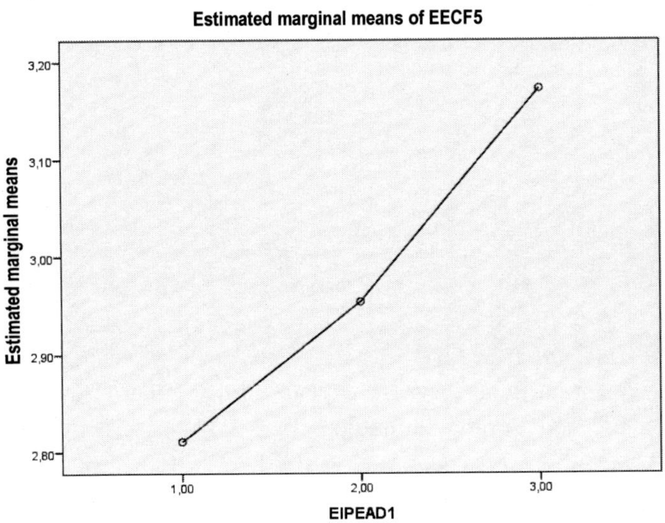

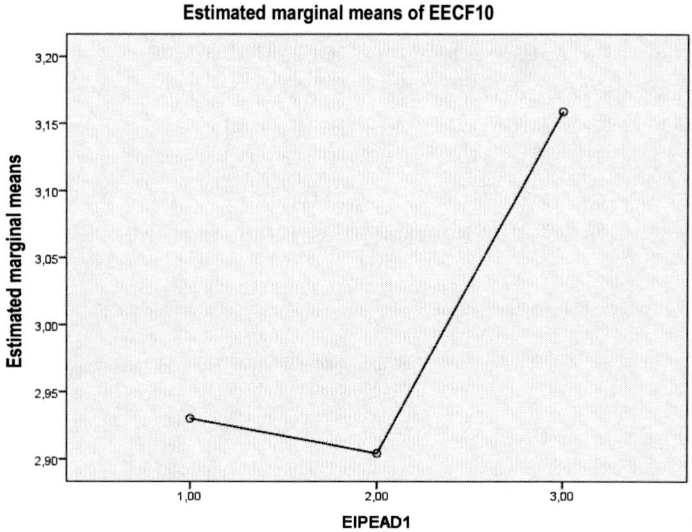

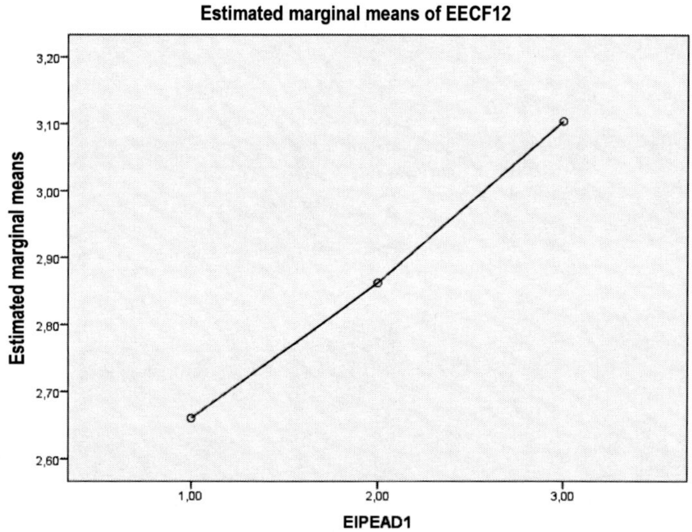

2) Emotion-focused strategies

A significant partial effect also appeared for regulatory teaching on factors of emotion-focused coping strategies: EEC9 [$F(2,165) = 2.946$, $p < .05$, $eta^2 = .934$, $3 < 1$, $p < .05$] and EECF11[$F(2,1658) = 5.514$, $p < .010$, $eta^2 = .063$, $3 < 1$, $p < .001$; $3 > 2$, $p < .05$], in an inverse sense (Figure 3).

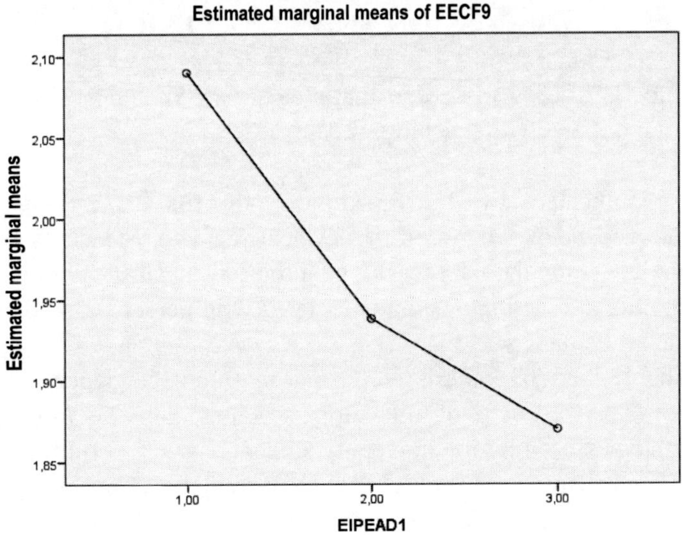

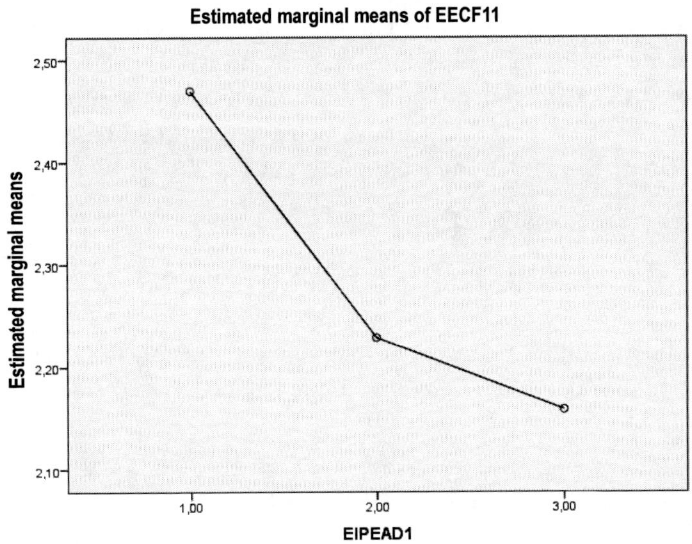

Note: 1 = Low level, 2 = Medium level; 3 = High level; EECF2 = Family help and counsel; EECF5 = Self-instructions; EECF10 = Positive reappraisal and firmness; EECF12 = Social support; EECF9 = Emotional venting, isolation; EECF11= Resigned acceptance.

Figure 3. Estimated marginal means. Levels of the coping strategy factors (EEC), according to the level of regulatory teaching (EIPEAD1).

Discussion and Conclusion

Evidence is offered to defend that *Regulatory Teaching* is an important characteristic of Effective Teaching, because:

1) As indicated by fulfillment of our first hypothesis, that of interdependence between the level of regulatory teaching and that of anxiety, regulatory teaching protects against strong negative emotional reactions, being associated with perceived predictability in the academic context.
2) The second hypothesis, referring to interdependence relationships between the level of regulatory teaching and the type of coping strategies used by the university students, was also fulfilled, although more decisively for problem-focused strategies. This indicates that the use of coping strategies, as a meta-emotional skill (de la Fuente, 2015; de la Fuente et al., 2014, 2015) is interdependent with the type of regulatory context. Regulatory teaching encourages security and predictability in the teaching-learning process, leading to lower levels of stress. Consequently, the student focuses on the problem, and does not need to focus on negative achievement emotions (De la Fuente, García-Torrecillas and Rodríguez-Vargas, 2015).

Practical Implications

With this result, we can focus intervention strategies on the design and development of *Regulatory Teaching*, in the university academic context.

The Design of Learning Environments

It is the teacher's responsibility to design learning environments that promote health and a proper emotional outlook. In this direction, there is sufficient evidence on how to prevent maladaptive emotions. Evidence-based recommendations are required in order to implement these (Roehrig et al., 2012).

1) *Cognitive quality of the instruction task.* The cognitive quality of the instruction increases students' perceived control and positive achievement emotions. Improving certain aspects of instruction such as clarity and structure can contribute to students' perception of control. From a cognitive perspective, materials that produce cognitive activation are preferable to those that promote routine procedures. It is also important how demanding the tasks are, because (1) they determine the difficulty of the material, and (2) this may influence the material's perceived value (Pekrun, 2006). Epistemic emotions should also be induced, through stimulating change in cognitive schemata, curiosity, productive confusion (Craig et al., 2008).
2) *Quality of motivation during the instruction.* The control-value theory (Pekrun, 2006) establishes that attribution of positive value is associated with academic engagement and performance. Teachers and peers provide direct verbal messages about academic value. There are two important ways to encourage attribution of high value: 1) By matching instructional material, task completion and interaction to the needs of the students, using authentic learning tasks; 2) Emotions that catch on from teachers to students, facilitating the material being absorbed, and the experience of positive values and emotions (Frenzel et al., 2009).
3) *Support for autonomy through teaching self-regulation.* When learning environments provide the opportunity for self-regulated learning, students develop self-regulation abilities. On one hand, this responds to students' need for autonomy, increasing their perceived control. On the other hand, self-regulated learning involves effective emotionality, involving self-direction competencies (Buff, Reuser, Rakoczy and Pauli, 2011). Along these lines, the DEDEPRO model has been proposed, regulating teaching for the purpose of self-regulated learning (De la Fuente and Justicia, 2007; De la Fuente et al., 2012; De la Fuente, García-Berbén and Zapata, 2013).
4) *Goal structures, practices and performance expectations.* Class goal structures define opportunities for success or failure. At the individual level, these may be learning goals, performance goals, or a combination (De la Fuente, 2004). The existence of different goals gives rise to different experiences of learning, perceived control and emotions. One important consequence is that teachers should avoid

making social or normative performance comparisons, due to the potential negative impact on students' emotionality. Perceived control and positive emotionality should be promoted, based on positive expectations.

5) *Design of tests and quizzes.* There is ample empirical evidence that test structure can be a determinant in test anxiety. Research on test anxiety has established that increased perceived control and decreased importance given to failure are associated with a decline in performance anxiety. Tests designed to diminish anxiety ought to: 1) have a clear structure of materials and procedure for study; 2) avoid excessive task demands; 3) include break time; 4) allow students a second assessment opportunity; 5) allow access to lecture notes, texts and computer; 6) use questions that activate working memory. All this is especially relevant for the more disadvantaged students.
6) *Performance consequences.* Depending on the consequences, we find an increase in anxiety, in value attribution, and in negative emotions with regard to future assessment events. Perceived control based on effort is associated with positive emotions and a reduction in negative emotions. By contrast, when there seems to be a disconnect between students' effort and their outcomes, perceived control is decreased and there is a rise in negative emotions, anxiety and hopelessness.

Treatment of Emotions

Some students experience an excess of negative emotions or a lack of positive ones, where these cannot be regulated by teachers, parents or peers. In these cases, psychological therapy for achievement emotions is recommended. There is limited evidence with respect to therapy for test anxiety, but existing evidence shows that it can be treated. The emotional dimension of anxiety can be treated with relaxation techniques and systematic desensitization. The cognitive dimension can be treated with multimodal therapies that encourage engagement and academic enjoyment, such as cognitive control, values and a thinking style of self-value, attention training, cognitive restructuring therapy or stress inoculation training (De la Fuente, Franco and Mañas, 2010).

ACKNOWLEDGMENTS

This research was funded through R&D Project, ref. EDU2011-24805 (2012-2015) of the Ministry of Science and Innovation (Spain), and with Federal Funds (European Union). http://www.estres.investigacion-psicopedagogica.com/english/seccion.php?idseccion=1.

REFERENCES

Acee, T. W., Kim, H., Kim, H. J., Kim, J., Hsieang-Ning, R. C. and Kim, M. The Boredom Research Group (2010). Academic boredom in under and overchallenging situations. *Contemporary Educational Psychology, 35*, 17-27.

Buff, A., Reusser, K., Rakoczy, K. and Pauli, C. (2011) Activating positive affective experiences in the classroom: "Nice to have" or something more? *Learning and Instruction, 21*, 452-466.

Craig, S. D., D'Mello, S., Witherspoon, A. and Graesser, A. (2008). Emote aloud during learning with Auto Tutor: Applying the Facial Action Coding System to cognitive-affective states during learning. *Cognition and Emotion*, *22*, 777-788.

Chorot, P. and Sandín, B. (1987). *Coping Strategies Scale (EEC).* Madrid: UNED.

Corno, L. (2008). On teaching adaptatively. *Educational Psychologist, 43*, 161-173.

De la Fuente, J. (2004). Recent perspectives in the study motivation: Goal Orientation Theory. *Electronic Journal of Research in Educational Psychology*, *2*(1), 35-62.

De la Fuente, J. (2011). Implications for the DEDEPRO Model for Interactive Analysis of the Teaching-Learning Process in Higher Education. In R. Teixeira (Ed.), *Higher Education in a State of Crisis* (pp. 205-222). New York: Nova Science Publisher Inc.

De la Fuente, (2014). *Confirmatory Structural Equation Model of Coping Strategies Scale (EEC).* University of Almería.

De la Fuente, J. (2015). *Competence for studying, learning and performing under stress: Self-help Guide for University Students, Graduates and Professional Examination Candidates.* Almería: Education and Psychology I+D+I, e-Publishing Series I+D+I©.

De la Fuente, J., Franco, C. and y Mañas, I. (2010). Efectos de un programa de entrenamiento en conciencia plena (midfullness) en el estado emocional de estudiantes universitarios [Efectos de un programa de entrenamiento en conciencia plena (midfullness) en el estado emocional de estudiantes universitarios]. *Estudios sobre Educación*, *19*, 31-52.

De la Fuente, J., García-Berbén, A. B. and y Zapata, L. (2013). How Development of Regulated Teaching Impacts Perceptions of University Students: the role of teacher training. *Infancia y Aprendizaje*, *36*(3), 375-378.

De la Fuente, J., García-Torrecillas, J. M. and Rodríguez-Vargas, S. (2015). The relationship between coping strategies, test anxiety, and bournout-engagement behavior in university undergraduates. In M. Bachman (Ed.), *Coping Strategies and Health*, (pp. 27-44). New York: NOVA Publisher.

De la Fuente, J. and Justicia, F. (2007). The DEDEPRO model for regulating teaching and learning: recent advances. *Electronic Journal of Research in Educational Psychology*, *13*(5), 535-564.

De la Fuente, J., López, M., Zapata, L. Martínez-Vicente, J. M., Vera, M. M., Solinas, G. and Fadda, S. (2014). Competence to Study and Learn in Stressful Contexts: Fundamentals of the e-Coping With Academic StressTM Utility. *Electronic Journal of Research in Educational Psychology*, *12*(3), 717-746.

De la Fuente, J., López, M., Zapata, L., Sollinas, G. and Fadda, S. (2015). Improving mental Health trough and online self-assessment and self-help e-Utility in university Students. In Roberta V. Nata (Ed.), *Progress in Education*, (vol. *33*, pp. 63-74), New York: NOVA PUBLISHER.

De la Fuente, J. and Martínez-Vicente, J. M. (2007). *Scales for Interactive Assessment of the Teaching-Learning Process, IATLP.* Almería, Spain: Education and Psychology I+D+i, e-Publishing Series I+D+i®.

De la Fuente, J., Zapata, L., Martínez-vicente, J. M., Cardelle-Elawar, M., Sander, P., Justicia, F., Pichardo, M. C. and García-Berbén, A. B. (2012). Regulatory Teaching and Self-Regulated Learning in College Students: Confirmatory Validation Study of the IATLP Scales. *Electronic Journal of Research in Educational Psychology*, *10*(2), 839-866.

Entwistle, N. J. and Entwistle, A. (1991). Contrasting forms of understanding for degree examinations: The student experience and its implications. *Higher Education*, *22*, 205-227.

Frenzel, A. C., Goetz, T., Lüdtke, O., Pekrun, R. and Sutton, R. E. (2009). Emotional transmission in the classroom: Exploring the relationships between teacher and student enjoyment. *Journal of Educational Psychology*, *101*, 705-716.

Friedricks, J. A., Blumenfeld, P. C. and Paris, A. H. (2004). School engagement: Potential of the concept, state of the evidence. *Review of Educational Research*, *74*, 59-109.

Goe, L., Bell, C. and Little, O. (2008). Approaches to evaluating teacher effectiveness: A research synthesis. Washington, DC: National Comprehensive Center for Teacher Quality. Retrieved October 15, 2015, from http://www.tqsource.org/publications/EvaluatingTeachEffectiveness.pdf.

Guthrie, J. T., McRae, A. and Klauda, S. L. (2007). Contributions of Concept-Oriented Reading Instructions to knowledge about interventions for motivations in reading. *Educational Psychologist*, *42*, 237-250.

Lazarus, R. S. and Folkman, S. (1984). *Stress, appraisal and coping.* New York: Springer Publishing Company.

Pekrun, R. (1992). The impact of emotions on learning and achievement. Towards a theory of cognitive motivational mediators. *Applied Psychology*, *41*, 359-376.

Pekrun, R. (2006). The control-value theory of achievement emotions: Assumptions, corollaries, and implications for educational research and practice. *Educational Psychology Review*, *18*, 315-341.

Pekrun, R. (2009). Global and local perspectives on human affect: implications of the control-value theory of achievement emotions. In M. Wosnitza, S. A. Karabenick, A. Eflides and P. Nenninger (Eds.), *Contemporary Motivational Research: From global to local perspectives*, (pp. 97-115). Toronto, Ontario, Canada: Hogrefe.

Pekrun, R. and Stephens, E. J. (2012). Academic emotions. In K.R. Harris, S. Graham and T. Urdan (EDs.), *APA Educational Psychology Handbook*, (vol. *2*, pp. 3-32). Washington, DC.

Reeve, J., Jang, H., Carrell, D., Jeon, S. and Barch, J. (2004). Enhancing students´s engagement by incresting teachers´authonomy support. *Motivation and Emotion*, *28*, 147-169.

Roehring, A. D. and Chistesen, E. (2010). Development and use of tool for evaluating effectiveness in grades K-12. In V. Schuter and B. Becker (Eds.), *Innovative assessment for the 21st century: Supporting educational assessment*, (pp, 207-228). New York, NY: Springer.

Roehrig, A. D., Turner, J. E., Arrastia, M. C., Christesen, E., McElhaney, S. and Jakiel, L. M. (2012). Effective teachers and teaching: Characteristics and practices related to positive student outcomes. In K.A. Harris, S.

Graham and T. Urdan (Eds.), *APA Educational Psychology Handbook*, vol. *2*, (pp. 501-527). Washington: American Psychological Association.

Spielberger, C. D. (1980). Preliminary Professional Manual for the Test Anxiety Inventory (Test Anxiety Inventory) T.A.I. Palo Alto, CA: Consulting Press.

Zimmerman, B. J. (1989). A social cognitive view of self-regulation academic learning. *Journal of Educational Psychology*, *81*, 329-339.

In: Teaching and Learning
Editor: Malcolm Vargas
ISBN: 978-1-63485-228-9

Chapter 7

INVERSE MODELING PROBLEMS AND THEIR POTENTIAL IN MATHEMATICS EDUCATION

Victor Martinez-Luaces*
Foundation J. Ricaldoni, Faculty of Engineering, UdelaR,
Montevideo, Uruguay

ABSTRACT

The traditional mathematics courses are dominated by routine exercises that require memory skills and, at best, they appear mixed with a few modeling problems which are always direct problems. However, real life problems are quite the opposite of traditional courses exercises: they are frequently inverse problems and also tend to require mathematical modeling skills. These problems will be called *inverse modeling problems*; they were studied in previous Mathematics Education papers, and in Applied Mathematics book chapters. In this chapter, inverse modeling will be analyzed first from a theoretical viewpoint followed by several detailed examples. During the last two decades, some of these problems were widely used at the University of the Republic of Uruguay (UdelaR), in engineering courses. Recently, several teacher training courses and workshops – in Argentina, Guatemala and Mexico – focused on Calculus inverse modeling problems were conducted with prospective and in-service teachers. The results of

* Corresponding Author address: Email: victorml@fing.edu.uy.

these experiences will be analyzed and finally, conclusions about the potential and richness of inverse modeling problems in Mathematical Education will be presented.

Keywords: inverse problems, modeling and applications, undergraduate courses, teacher training courses

INTRODUCTION

The traditional mathematics courses are dominated by routine exercises that require memory skills and, at best, appear mixed with a few direct modeling problems. However, real life problems are frequently inverse problems that tend to require mathematical modeling skills. These are in fact the opposite of exercises found in traditional courses. These problems will be called *inverse modeling problems* and they have been studied previously in Mathematics Education papers (Martinez-Luaces, V., 2007, 2009a, 2009b, 2009c and 2013), and in Applied Mathematics book chapters (Martinez-Luaces, V., 2012a, 2012b and 2015).

According to Groestch (Groestch, C.W., 1999 and 2001), direct problems are those that provide the required information in order to follow a well-defined stable procedure leading to a single correct solution. An example of a possible inverse problem consists in supplying the students with the output and then, asking them to obtain a possible input that fits with that particular output; this is referred to as a *causation inverse problem* (e.g., provide a set of functions and ask for an ordinary differential equation (O.D.E.), where the solutions are the given functions). Finally, if both input and output are known, the question is about the middle process, then, it is known as *specification inverse problem* (e.g., provide hypothesis and thesis of a mathematical proposition and ask for a possible proof).

Both of these types of inverse problems are usually more difficult and interesting than the corresponding direct ones, largely due to either multiple solutions or no solution (Bunge, M, 2006).

For example, if n and m are prime numbers, where $n \neq 2$ and $m \neq 2$, it follows that $s = n + m \geq 6$ is an even number, whereas the inverse problem of the decomposition of an even number to a sum of a couple of prime numbers is extremely difficult. Moreover, this problem is related to the even Goldbach conjecture, one of the oldest and best-known unsolved problems in number theory and in all of mathematics (Oliveira e Silva, et al., 2014).

Inverse problems are common in experimental sciences textbooks. For instance, all modern organic chemistry textbooks include an important set of organic synthesis (Vollhardt, K. et al., 2007). This kind of problem is solved by Retrosynthetic Analysis, a technique developed by E.J. Corey, who was awarded with the Nobel Prize in Chemistry 1990 (Corey, E.J., 1991).

In Mathematical Education the situation is completely different, as inverse problems are almost ignored. Moreover, if inverse problems do require modeling skills in order to be solved – thus mathematically enriching them – their presence in mathematics courses and textbooks is rare and similarly, hardly seen in assessment tasks. For this reason, studies of inverse problems and problems that require modeling skills have been previously investigated in Uruguayan traditional Calculus and pre-Calculus courses (Martinez-Luaces, V., 2011a). The results of this study have shown that only a few problems could be considered inverse and/or modeling problems and only a few had both traits. The current set of problems may not be enough to prepare students for professional and real-life situations.

In this chapter, inverse modeling will be analyzed first from a theoretical viewpoint followed by several detailed examples. These examples will include several simple Calculus problems, a Linear Algebra inverse modeling problem solvable by two different techniques and finally some difficult problems modeled through linear ODE systems.

During the last two decades, some of these problems – in simplified versions – were widely used at the University of the Republic of Uruguay (UdelaR), in engineering courses. Recently, several teacher training courses and workshops took place in Tucuman (Argentina, 2009), Guatemala (Guatemala, 2010) and Colima (Mexico, 2013). These courses were designed for prospective or in-service teachers and they focused on Calculus inverse modeling problems.

The results of these experiences will be analyzed and finally, conclusions about the potential and richness of inverse modeling problems in Mathematical Education will be presented.

THEORETICAL FRAMEWORK

Modeling and Applications

In the preliminary discussion document to the ICMI (International Commission on Mathematical Instruction) Study 14 (Blum, W., 2002), the term "modeling" focuses on the direction that goes from the real world towards mathematics, while the term "applications" implies the opposite direction. In addition, the term "modeling" emphasizes the process that is taking place, while the word "applications" stresses the object involved, particularly real life examples that are susceptible to mathematical manipulation.

Therefore, we can arrive at the following schema:

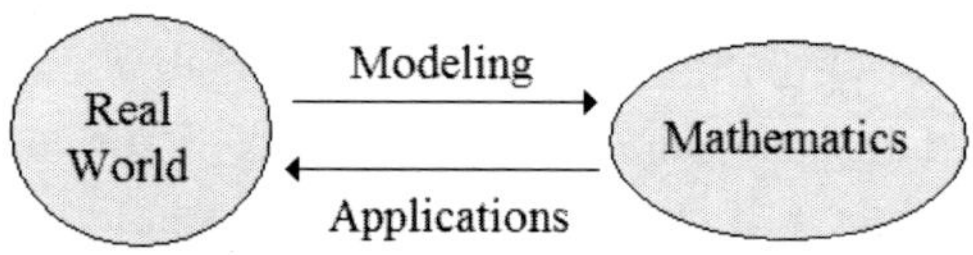

Figure 1. Scheme of modeling and applications.

An extended discussion about modeling and applications in Latin American countries can be found in papers (Martinez-Luaces, V., 2005 and 2009a).

Inverse Problems

According to Groestch (Groestch, C.W., 1999 and 2001), direct problems are those that provide the necessary information to follow a well-defined, stable procedure leading to a single correct solution. This process can be schematized as follows:

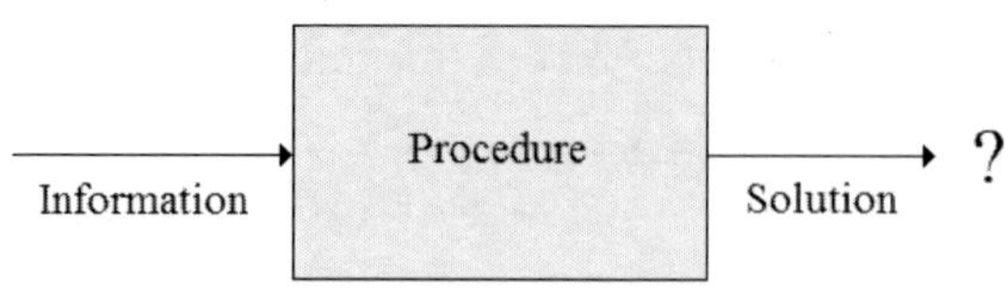

Figure 2. Scheme for direct problems.

In contrast, inverse problems may have multiple solutions or simply no solutions, thus making them more interesting though consequently more difficult (Bunge, M., 2006). In essence there are two types of inverse problems; firstly, the causation problem, where the procedure is well-known and the question is concerned with the necessary data in order to obtain a given result. This situation is schematized in Figure 3:

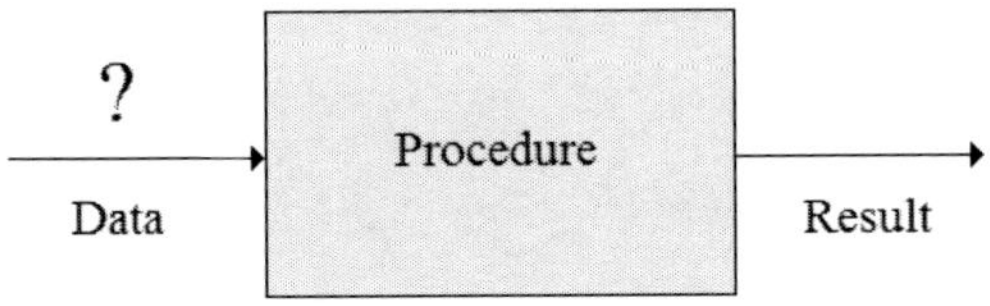

Figure 3. Scheme for inverse causation problems.

The other inverse problem encountered is the specification problem, where data and result are given and the question is concerned with which procedure can give the desired result (output) with the given data (input). This process is schematized in Figure 4:

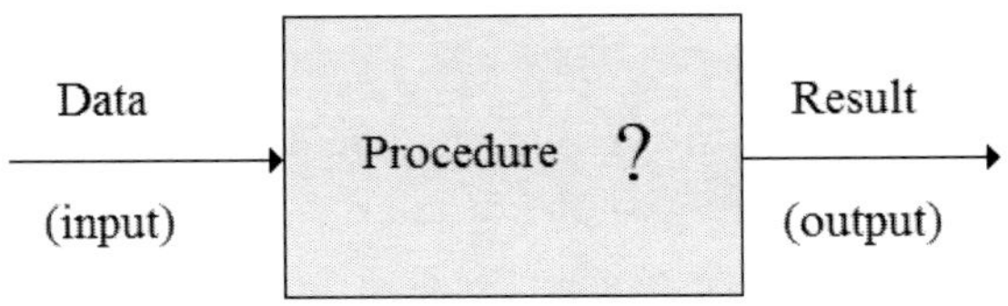

Figure 4. Scheme for inverse specification problems.

Both of these problems are common in the experimental sciences and real life situations, as noted in previous research (Martinez-Luaces, V., 2011b and 2011c).

THREE SIMPLE CALCULUS EXAMPLES

A Problem about Areas

In this problem, a sheep is eating grass in a square field with side length L. The sheep is tied at the point $(L/2\ ,\ 0)$ and the rope attached to the sheep has a length R as shown in Figure 5:

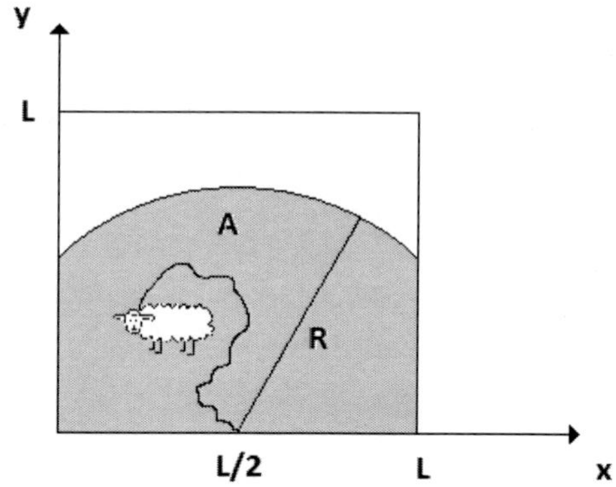

Figure 5. Part of the field accessible for the sheep.

In Figure 5, A represents the area of the sector where the sheep may graze. Let $r=\frac{R}{L}$ be the ratio of the rope length to field side length and let $f=\frac{A}{L^2}$ represent the fraction of the total area which is accessible for the sheep. Since f is a function of the ratio r, then the solution can be obtained using integration.

Typical exercises consist in supplying students with this figure and asking them to obtain f corresponding to one or more values for r. However, a more interesting approach is to ask the students to draw a diagram hoping they realize that the problem can be solved as an intersection of circles and squares. Four different situations may be occurring:

1. When $r\in[0,1/2)$, the sheep does not reach the lateral edges of the field.
2. When $r\in[1/2,1)$, the sheep can reach the lateral edges, but not the upper one.
3. When $r\in\left[1,\frac{\sqrt{5}}{2}\right)$, the sheep can reach the top edge of the field, but not the whole field.
4. When $r\geq\frac{\sqrt{5}}{2}$, the sheep can graze all around the field.

This problem requires modeling and integral calculation and it can be easily converted into an inverse problem. It is obvious that for every value of $r \geq 0$ there exists a unique value of A, but more challenging is to ask the question from another angle. For example, for any value of A, does a corresponding unique value of r exist or not? To solve this problem, the function $f(r)$ must be studied in terms of continuity and growth with $r \geq 0$, in order to ensure its invertibility.

A Problem about a Sand Hopper

A sand hopper is discharging sand at a rate of $0.4\, m^3/s$ forming a conical pile so that the radius of the base and height are the same, as it is shown in the following figure:

Figure 6. Discharge sand from the sand hopper.

The typical direct problem consists in determining the speed of the increasing height of the sand pile, for example, when its volume is $1.0\, m^3$. This modeling problem may be inverted in several ways, for instance asking how many cubic meters per minute of sand are discharged, when given the height of the pile of sand and the total time of discharge. Another option consist in providing the rate of change of the radius and asking how long the hopper had been working.

This example shows how a direct modeling problem can be converted into different inverse modeling problems and thus adding depth to the question.

A Problem about Doughnuts

A doughnuts manufacturer produces a standard size, having an inner radius of $2.0\ cm$ and an outer radius of $4.0\ cm$. He also wants to produce the doughnuts with a chocolate coating and colored chips on the upper half, as shown in Figure 7.

Figure 7. Doughnut with chocolate and colored chips covering.

Obviously, the manufacturer knows the prices of dough and chocolate (in $\$/m^3$) and the cost of the colored chips topping (in $\$/m^2$). A typical direct problem consists of calculating the unit cost of the doughnuts, and then determine manufacturing costs and estimate of a possible sale price per unit and the corresponding company profits.

It is important to note that in order to solve this problem, it is enough to have a previous knowledge of the torus area and volume formulae – or how to obtain them using integration methods – and then calculate the areas and volumes involved and multiply by the corresponding costs of dough, chocolate and colored chips.

For instance, if the doughnut is modeled as a circular torus, its corresponding volume is $V = \frac{1}{4}\pi^2(r+R)(R-r)^2$ and the corresponding surface area is $A = \pi^2(R^2 - r^2)$ where r and R represent the inside and outside radius, respectively.

This example can easily be converted into an inverse modeling problem just by fixing the unit price and choosing values for the thickness of the chocolate covering, the area to be covered with the chips, the internal and the external radius, etc. For instance, the inner radius ($2\ cm$) and the outer radius ($4\ cm$) could be the same as in the direct problem, and the inverse problem could consist in obtaining the best values for the chocolate thickness and the area covered in chips.

As these examples show, it is not difficult to transform a modeling problem into an inverse one. This process enriches the problem and allows students to explore existence, uniqueness and stability that are not so interesting in direct problems (as these properties are often trivial).

As a consequence of these facts, a greater understanding of the problem is needed for solving it when is presented in its inverse version.

A LINEAR ALGEBRA EXAMPLE IN TWO DIFFERENT VERSIONS

Linear algebra may be considered as a second level abstraction (Guzman Ozamiz, M. 2010), and as a consequence, modeling problems are not frequently found as easily as in Calculus or Differential Equations.

On the other hand, as it was mentioned before, the undergraduate curriculum is dominated by direct problems and Linear Algebra courses are not an exception to this rule.

As a result of these two facts, it is not easy to get inverse modeling problems in this branch of mathematics.

In this section a Linear Algebra problem – which is an example of inverse modeling problem – will be analyzed. This problem will be re-formulated from the corresponding direct problem and will be solved using two different approaches: one of them based on linear transformations and the other using transformations in the complex plane.

Here is the direct problem: The pirate Knight – who is a chess fan – goes to an island with the intention of hiding treasure. The island has two large trees and the pirate decides to put his flag, at a certain distance from those trees. Then, the pirate proceeds to follow the movements of the chess knight: he walks straight to the first tree, walking a distance L_1 and then, he turns 90 degrees clockwise and he walks a distance $2L_1$ and finally he puts a stick at this point (labeled as S_1 in Figure 8a). After that, he repeats the procedure with respect to the second tree, except that he turns 90 degrees counter clockwise and after completing the process he puts a second stick (labeled as S_2 in Figure 8b). The pirate's treasure is buried at the midpoint M, between the sticks S_1 and S_2. The direct problem consists in finding the point where the pirate buried his treasure as a function of the positions of the trees and the flag.

Figures 8a and 8b show a diagram corresponding to the pirate's movements, in order to arrive to the points S_1 and S_2, which are the positions of the sticks.

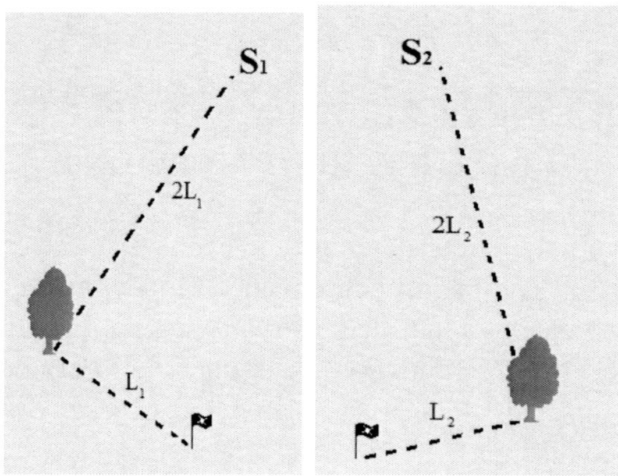

(a) (b)

Figures 8a and 8b. Flag, trees and sticks positions.

This direct problem can be converted into a more interesting inverse question, as follows: suppose that the pirate returns to the island years later and finds that the sticks and flag have disappeared. Then, the inverse problem to solve is: where could the flag be placed in order to find the treasure under the new conditions in which the pirate finds the island, i.e., with standing trees, but no flag or sticks?

The Linear Transformations Approach

If the coordinates system is chosen such that the tree on the left is positioned at the origin and both trees lie on the x axis, then the tree positions will be $(0,0)$ and $(L,0)$ respectively, being L the distance between both trees.

This situation could be easily represented as in Figure 9.

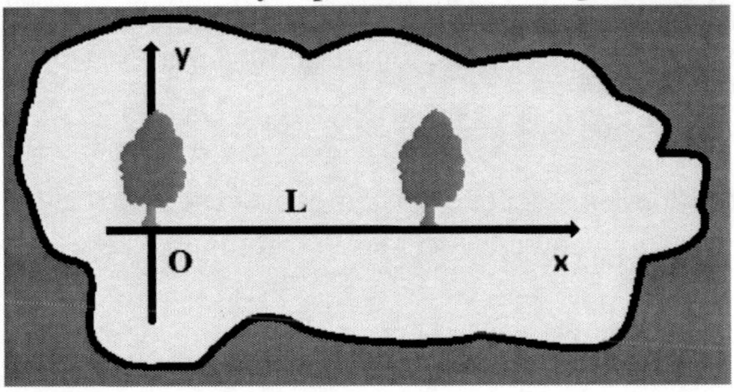

Figure 9. The pirate island and the trees.

Let us suppose that the flag was somewhere in the southern part of the island, for instance, at the point $(a,-b)$ as in Figure 10.

Two important vectors, $\vec{u}=(0,0)-(a,-b)=(-a,b)$ and $\vec{v}=(L,0)-(a,-b)=(L-a,b)$ have been included in Figure 10.

If R_1 is the first rotation (i.e., the 90 degrees clockwise rotation), then $R_1(\vec{u})=\begin{pmatrix}0 & 1\\ -1 & 0\end{pmatrix}\begin{pmatrix}-a\\ b\end{pmatrix}=\begin{pmatrix}b\\ a\end{pmatrix}$, and the first stick will be located at the point $(2b,2a)$. Similarly, if R_2 is the linear transformation corresponding to the 90 degrees counterclockwise rotation, then the rotated vector, is $R_2(\vec{v})=\begin{pmatrix}0 & -1\\ 1 & 0\end{pmatrix}\begin{pmatrix}L-a\\ b\end{pmatrix}=\begin{pmatrix}-b\\ L-a\end{pmatrix}$. If this vector is multiplied by 2 and moved such that its point of application matches the tree on the right, located at $(L,0)$, then $2(-b,L-a)+(L,0)=(L-2b,2L-2a)$ will be the point for the second stick.

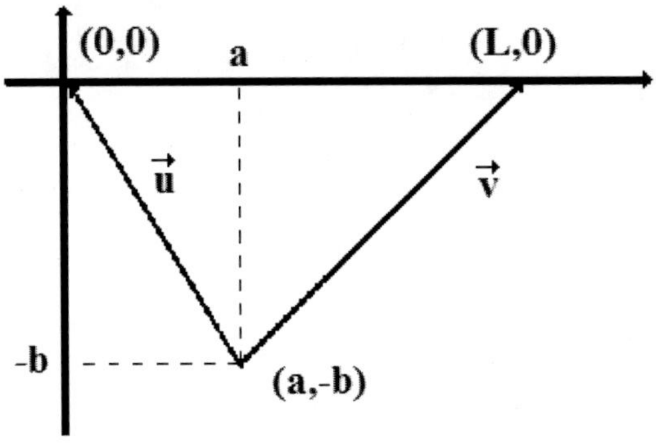

Figure 10. Relevant points and vectors in the pirate problem.

Finally the treasure would be found at the midpoint M, which can be easily obtained as: $M = \frac{1}{2}(2b,2a) + \frac{1}{2}(L-2b,2L-2a) = \left(\frac{L}{2},L\right)$.

As this result does not depend on the coordinates of the flag position, it is obvious that the treasure can be found starting from anywhere in the southern part of the island.

Repeating the procedure for a point on the north side of the island, the same result is obtained. Interestingly when a group of students worked on a different version of this problem, they created a computer animated presentation to show this result, by choosing different points for the flag in both southern and northern parts of the island.

As a consequence, the position of the flag and the sticks are not relevant to the location of the buried treasure.

The Second Approach Based on Complex Numbers

In this second approach, let us think about putting the real axis on the line connecting the trees and the imaginary axis will be the perpendicular bisector (see Figure 11).

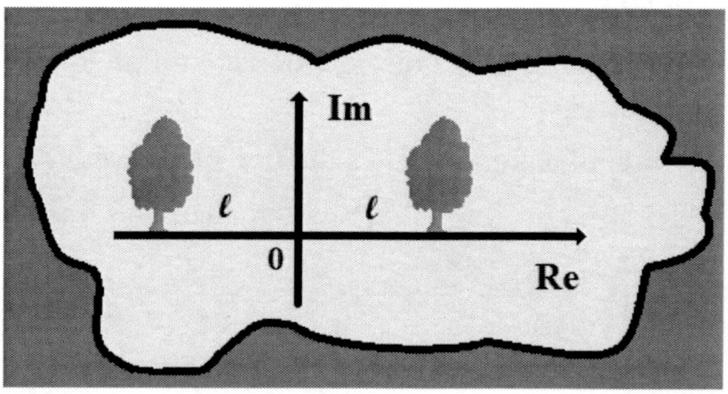

Figure 11. The pirate island and the complex plane.

In this case the trees' positions are $-l+i0$ and $l+i0$, while the pirate flag is located in $a+ib$, which is a generic point in the southern part of the island. The positions are indicated in Figure 12.

Remembering that the pirate proceeds to follow the movements of the knight as in the chess game, if $\vec{z}_1 = -l + i0 - (a+ib) = (-l-a) - ib$ is the vector that joins the flag and the tree on the left, then the clockwise rotation gives a new vector $-i\vec{z}_1 = -b + i(l + a)$ and finally, the first stick will be located at $-l + i0 - 2i\vec{z}_1 = (-l - 2b) + i(2l + 2a)$.

Now, the same process is carried out to find the position of the second stick. In this case $\vec{z}_2 = l + i0 - (a + ib) = (l - a) - ib$ is the vector that joins the flag and the tree on the right, the counterclockwise rotation gives the vector $i\vec{z}_2 = b + i(l - a)$ and the second stick will be located at $l + i0 + 2i\vec{z}_2 = (l + 2b) + i(2l - 2a)$.

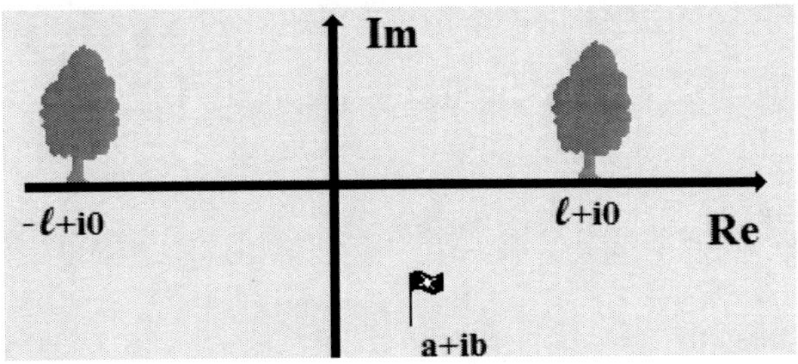

Figure 12. Flag and trees positions.

Finally, the treasure will be in the midpoint M of the sticks, i.e., $M = \frac{1}{2}[(-l - 2b) + i(2l + 2a)] + \frac{1}{2}[(l + 2b) + i(2l - 2a)] = 0 + i2l$, which is the same point obtained in section 4.1, but with a different notation.

O. D. E. and O. D. E. Systems Inverse Modeling Problems

Inverse Problems in Chemical Reactions Mathematical Models

A typical direct problem that involves first order O.D.E. appears in the mathematical model of a unimolecular chemical reaction. This separable variables O.D.E. problem is usually included in classical mathematics textbooks such as Courant (Courant, R., 1937).

The original problem comes from an important paper published by L.E. Wilhelmy in 1850 (Wilhelmy, L., 1850). The German physicist studied the inversion of sugar with acids, using a new technique (Polarimetry) for evaluating the dependency of reaction velocity on the quantity of reactants and temperature. As S. Zambelli mentioned, "…in this paper probably appeared the first differential equation in chemistry" and "…although written in a prestigious journal the paper passed unnoticed by contemporary scholars. It will be rediscovered only in 1884 by Ostwald" (Zambelli, S., 2012).

In a simplified version, this unimolecular first order chemical reaction can be easily schematized as follows:

$$A \xrightarrow{k} B$$

where A represents sugar, k is the kinetic constant and B is the inverted sugar.

The corresponding O. D. E. mathematical model is:

$$\frac{d[A]}{dt} = -k[A]$$

where $[A]$ represents the concentration of substance, t is time and the negative sign indicates that the sugar is being transformed, so its concentration diminishes with time.

The O.D.E. solution is:

$$[A] = [A]_0 \exp(-kt)$$

where $[A]_0$ is the initial concentration of the reactant.

In both cases (Courant's textbook and Wilhelmy's paper), the authors were concerned about the sugar concentration and how it varies with time. Let us consider a different point of view that of the study of the stability of O.D.E. solutions. Before analyzing this new direct problem, it is important to mention that stability has an important chemical consequence, since it implies that small errors in the initial concentration measurements will remain bounded as time progresses. Moreover, if the O.D.E. solutions show an asymptotic stability, these errors will tend to disappear as $t \to +\infty$.

If we presume that the initial concentration changes from $[A]_0$ to $[A]_0^*$ then difference in the reactant concentrations will be:

$$|\ [A] - [A]^*\ | = |\ [A]_0 - [A]_0^*\ | \exp(-kt) \xrightarrow[t \to +\infty]{} 0$$

As a consequence, the O.D.E. solutions are asymptotically stable and small errors in the initial concentrations will tend to vanish as time progresses.

The following train of thought leads to an interesting inverse problem. Is it possible to propose a chemical reaction where solutions show a different kind of stability (i.e., instability or at least a weak stability)?

In order to answer this inverse problem, let us consider the situation that takes place when the reaction between A and B is reversible. In this case, we will have:

$$A \xrightarrow{K} B \text{ and } B \xrightarrow{k} A$$

where K and k are the corresponding kinetic constants. It is important to note that this is not a theoretical reaction, in fact a real-life example of this chemical reaction is given by the Mutarotation of Glucose (Guerasimov, Y. et al., 1995). This situation can be schematized more easily as:

$$A \underset{k}{\overset{K}{\rightleftarrows}} B$$

And the corresponding O.D.E. system will be:

$$\begin{cases} \dfrac{d[A]}{dt} = -K[A] + k[B] \\ \dfrac{d[B]}{dt} = K[A] - k[B] \end{cases}$$

This system can be written more easily as $\dot{X} = AX$, being

$$A = \begin{pmatrix} -K & k \\ K & -k \end{pmatrix}$$

Taking into account that the kinetic constants are always positive, the matrix eigenvalues are $\lambda_1 = -(K+k) < 0$ and $\lambda_2 = 0$ (Martinez-Luaces, V., 2012). As a consequence, the O.D.E. system solutions will show a weak stability, i.e., they will be stable but not asymptotically.

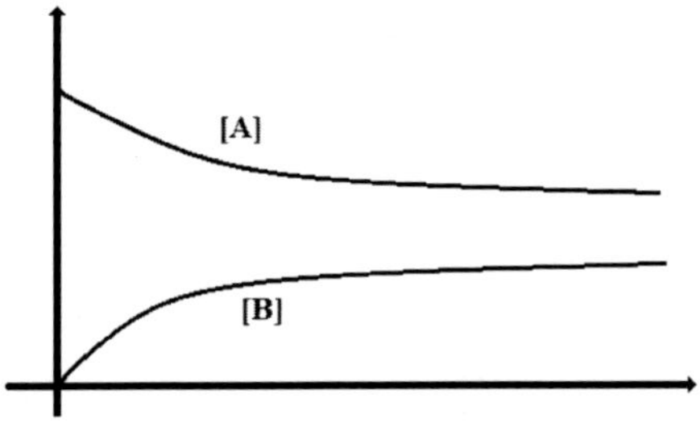

Figure 13. The concentrations of the chemical agents.

The corresponding concentrations can be graphically represented as in Figure 13.

As it can be seen in the previous graphics, all curves are monotonic functions. Now, this can be used to propose another inverse problem: is it possible to find a chemical mechanism where curves are not monotonic functions?

To answer this question we can analyze the following situation involving three different chemical species, where a chemical substance reacts to produce a chemical compound which reacts again to produce C , the final product of the sequence of reactions (Martinez-Luaces, V., 2005). This is a very common situation in real life, for example, when grape juice is converted into wine and then, it is transformed into vinegar.

This sequence of reactions is a first order chemical kinetics mechanism (F.O.C.K.M.), which can be represented as follows:

$$A \xrightarrow{k_1} B \xrightarrow{k_2} C$$

where k_1 and k_2 are the corresponding kinetic constants.

In the production of wine and vinegar, A is a carbohydrate, B is ethylic alcohol, and C is acetic acid. The mathematical model for this F.O.C.K.M. is the following:

$$\begin{cases} \dfrac{d[A]}{dt} = -k_1[A] \\ \dfrac{d[B]}{dt} = k_1[A] - k_2[B] \\ \dfrac{d[C]}{dt} = k_2[B] \end{cases}$$

and the associated matrix is:

$$M = \begin{pmatrix} -k_1 & 0 & 0 \\ k_1 & -k_2 & 0 \\ 0 & k_2 & 0 \end{pmatrix}$$

whose eigenvalues are obvious $\lambda_1 = -k_1 < 0$, $\lambda_2 = -k_2 < 0$ and $\lambda_3 = 0$ (note that the system matrix is a lower triangular matrix). If the corresponding eigenvectors are obtained, the solutions could be graphically represented as in Figure 14.

It can be shown that the inflection point of the curve of concentration $[C]$ occurs at the same time that the curve of concentration $[B]$ reaches its maximum. This can then be used to propose another inverse problem: is it possible to get curves with two or more inflection points in three chemical species F.O.C.K.M.s ?

The answer to this new inverse problem is negative. The corresponding analysis was shown in previous papers and book chapters where the curves of the adsorption of Carbon Dioxide (CO_2) on Platinum (Pt) surfaces were studied and then, the obtained results were generalized to any F.O.C.K.M. involving three chemical substances (Martinez-Luaces, V. 2012a), (Zinola, F. et al., 1997), (Martinez-Luaces, V. and Guineo, G., 2002).

It is important to mention that simplified versions of all those inverse problems were proposed in courses for chemical engineering and food technology students.

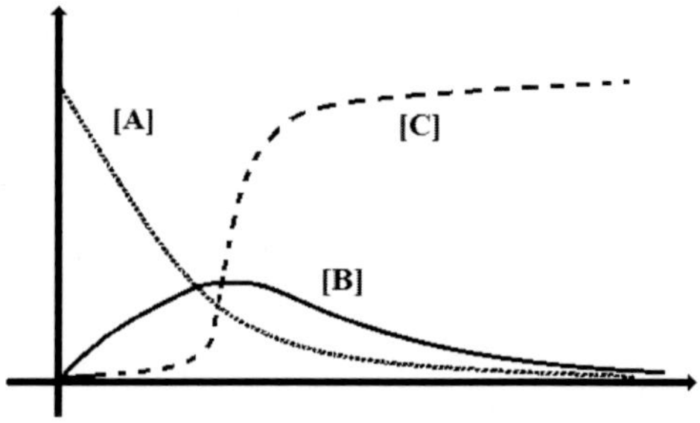

Figure 14. Sketches of curves of three chemical agents concentrations.

Some Examples of Difficult Inverse Modeling Problems

It can be noted that in all the previous cases, all the eigenvalues were negative, except one, which was zero.

It should be observed that all the previous chemical or electrochemical processes considered were quite different to each other; however all their mathematical models showed certain similarity. Particularly, in all cases $\lambda = 0$ was an eigenvalue of the O.D.E. system, so a more difficult inverse modeling problem could be if it is possible to propose a F.O.C.K.M. such that all its eigenvalues verify that $\lambda \neq 0$. It is important to observe that the answer to this question cannot be obtained just analyzing particular examples as in the previous cases.

For this purpose, the general form corresponding to a given F.O.C.K.M. must be studied. In a previous book chapter, published by NOVA in 2015 (Martinez-Luaces, V., 2015), the following general F.O.C.K.M. matrix was obtained:

$$A = \begin{pmatrix} -s_1 & k_{21} \cdots & k_{n1} \\ k_{12} & -s_2 \cdots & k_{n2} \\ \vdots & \vdots & \vdots \\ k_{1n} & k_{2n} \cdots & -s_n \end{pmatrix}$$

In this matrix all the non-diagonal entries are non-negative and the diagonal elements are $-s_i = -\sum_{j \neq i} k_{ij}$, so all the matrix columns add to zero.

Two obvious corollaries of this result can be easily obtained for this general form:

- $\det(A)=0$.
- $\lambda=0$ is an eigenvalue of A.

The first one is a straightforward conclusion from $row_1+row_2+\ldots+row_n=\vec{0}$ and the second one is an obvious consequence of having a null determinant.

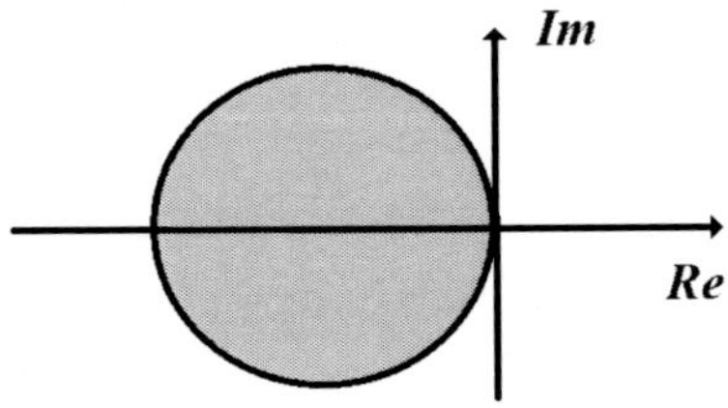

Figure 15. The Gershgorin Circle for a general F.O.C.K.M.

This analysis shows that if two or more substances are involved, it is impossible to propose a F.O.C.K.M. such that $\lambda \neq 0$, $\forall\, \lambda$, and the answer to the inverse problem is negative.

Another observation that can be made taking into account all the previous cases considered, is that all the eigenvalues were negative, except one, which was zero. Then, another interesting inverse modeling problem could be: Is it possible to propose a F.O.C.K.M. with a positive eigenvalue? Once again, the answer is negative but the proof is not obvious.

This result was proved using the Gershgorin Circle Theorem, a result that was first published by S. A. Gershgorin in 1931 and may be used to bind the spectrum of a complex $n \times n$ matrix (Varga, R., 2004).

If the Gershgorin Circle Theorem is applied to the F.O.C.K.M. matrix A, it can be proved that all the eigenvalues lie in a circle which is tangential to the imaginary axis, like the circle showed in Figure 15.

According to the previous results, it follows that:

$$\mathrm{Re}(\lambda_i) \leq 0 \;\; \forall i=1,2,\ldots,n \text{ and } \mathrm{Re}(\lambda_k)=0 \Leftrightarrow \lambda_k=0$$

This result proves that it is impossible to have an F.O.C.K.M. with a positive eigenvalue. Going back to the stability issues, it is important to note that the solutions corresponding to a non-zero eigenvalue λ are of the form:

$$\exp(\lambda t), t\exp(\lambda t), t^2 \exp(\lambda t), \ldots, t^p \exp(\lambda t),$$

which depend on the algebraic multiplicity and the corresponding geometric multiplicity of the eigenvalue λ. If $\mathrm{Re}(\lambda) < 0$, all the previous functions tend to vanish with time, independently of corresponding algebraic and geometric multiplicities.

The main problem – from a stability point of view – involves the multiplicities of the null eigenvalue. For this purpose it is easy to observe that the O.D.E. system solutions associated with the null eigenvalue are linear combinations of the following functions: $\{e^{0t}, te^{0t}, \ldots, t^q e^{0t}\}$, or the equivalent: $\{1, t, \ldots, t^q\}$. Then, the solutions due to the null eigenvalue are polynomial functions, where the degree q depends on both the algebraic and the geometric multiplicities corresponding to $\lambda = 0$.

It follows straightforward that only if $q = 0$, the polynomial solutions will remain bounded when t tends to infinity.

To sum up the previous results, it can be stated that only the null eigenvalue, in particular, its algebraic and geometric multiplicities are relevant to make predictions about the stability of the O.D.E. system solutions.

So, taking into account all the previous results, the final and most relevant inverse modeling problem will be if it is possible to find an unstable F.O.C.K.M. or equivalent, if it is possible to find a F.O.C.K.M. such that the algebraic and geometric multiplicities of the null eigenvalue are different. The answer to this question, once again, is negative as it was proved in a recent paper (Martinez-Luaces, V., 2016)

This final result shows that the O.D.E. solutions for any involved F.O.C.K.M. are always stable, but not asymptotically. This weak stability has an important chemical consequence, since it implies that small errors in the initial concentration measurements will remain bounded as time progresses, but they will not tend to disappear when $t \to +\infty$.

Other qualitative results can be obtained by analyzing the form of the solutions for the O.D.E. linear system. For instance, the existence and number of inflexion points in curves of concentration versus time were previously obtained in (Martinez-Luaces, V, 2012a), among other conclusions.

It is important to note that these problems – which are more difficult than those of the previous section – were used in postgraduate courses and/or in teaching training courses. For this purpose, simplified versions of these problems were presented in advanced courses with more mathematical guidance and advice offered to students.

Some of these experiences will be commented in the following section.

OUR EXPERIENCES

Experiences in Undergraduate Courses

Our first experience of teaching inverse problems and modeling started in 1996, in a course recommended for Food Technology Engineering students, followed by other courses tailored to Food Technology and Chemical Engineering students' needs. All students had a common experience of tackling inverse problems in other subjects as Analytical, Inorganic and Organic Chemistry. In contrast, they had virtually no previous experience of real-life modeling, as only some applications had been taught in previous courses at Secondary School. In Uruguay, secondary school teachers make the decision whether or not to include such applications and modeling problems in their courses depending on what connections to other disciplines is required.

As a result, mathematical modeling was taught from the first year of the course (i.e., 1996), in order to fill the gap in the students' knowledge. In subsequent years, inverse problems were included: causation problems in 1997, specification problems in 1998 and after 1999 all types of inverse problems (causation, specification and mixed) were studied, using modeling and mathematical applications skills and tasks, simultaneously.

Inverse Modeling Problems in Teacher Training Courses

As it was noted by Allan Schoenfeld, in the Fourth International Congress of Mathematical Education (ICME-4) in 1978, only one reference in one session was made to problem solving. Fortunately, a few years later, at ICME 5 in Australia (1984), the situation had changed dramatically (Schoenfeld, A.,

1994). In Latin American countries, this movement of mathematics education towards problem solving arrived at the beginning of the 1990s (Gonzalez, B., 2001), when textbooks in Spanish about problem solving appeared in the late 1980s (Mason, J. et al., 1988) and early 1990s (Guzman, M., 1991).

The same situation happened with modeling and application problems in Mathematics courses. The late arrival of this practice was not the only problem, for instance, in Uruguay, statistical studies showed that teachers in pure mathematics did not demonstrate real-life model problems in their classes and similarly at pedagogical faculties or tertiary institutes (Martinez-Luaces, V., 1998) (Gomez, A. and Martinez-Luaces, V., 2002). This situation was not just applicable to Uruguay but to other Latin American countries as well (Martinez-Luaces, V., 2005) (Gonzalez, B., 2001) (Hernandez, M., 1997). In fact, all the sub-continent had exactly the same problem, as mentioned in the Working Group on mathematical models at the Seventeenth Latin American Congress in Mathematical Education (RELME XVII), (Martinez-Luaces, V. et al., 2004)

If inverse problems are now examined, the situation could be considered the same or even worse (Bunge, M., 2006), (Martinez-Luaces, V., 2011b).

In an attempt to solve this problem, several courses, workshops, etc., for teachers and prospective teachers, were undertaken first in Uruguay and then in other Latin American countries. These courses and workshops were divided into three blocks:

- the first part presented theoretical concepts, mathematical tools, and typical examples of how to use them to model real-life problems
- in the second part, participants were encouraged to propose new problems related to their own teaching subjects or real-life situations
- these new problems (and the workshop itself) were considered in a plenary session, which represents the third and last part of the activities.

The first of these workshops to take place outside Uruguay was held at Tucuman, (Argentina, 2009) and it was a short course containing the topics O.D.E. and Laplace Transform. In the first part of this course, direct modeling problems were introduced and discussed with the participants (Argentinean university teachers) in terms of their richness and potential to be used in undergraduate courses. In the second part, inverse modeling problems (mainly causation problems) were analyzed, using the same ideas and mathematical tools.

The second course was held in Guatemala in 2010. Here, the course was devoted to inverse problems in general, not just inverse modeling problems. This course started with a theoretical framework, followed by a workshop about inverse problems and finally there was a general discussion about possible uses.

The third and last experience was in Colima (Mexico, 2013) where a ten hour mini-course and workshop was delivered. In this case, after a theoretical framework, the students and prospective teachers were faced with ten different direct modeling problems.

After this brief introduction, the course was transformed into a workshop, where the students – working in small groups – were expected to suggest inverse modeling problems based on the previously discussed direct problems. Finally, the different proposals were analyzed from the viewpoint of their mathematical education potential to be used in their classes.

RESULTS

Results Obtained in Undergraduate Courses

Questionnaires were administered to students towards the end of semester nearing final examinations. Answers were anonymous and no extra credit was awarded.

37 responses were collected, open-ended questions solicited their views about the courses, and particularly about inverse problems and mathematical modeling and applications.

Table 1 shows an overall impression, positive, negative or neutral with a brief description of the topic.

26 positive comments were divided into 4 sub-groups: general, usefulness, interaction and teaching style and the following selection highlight the general consensus of positive opinions.

General:

> "A really super course, I got a great deal out of it."
> "An interesting course, with quite a lot of applications in real life."

Table 1. Description of student's opinions

		Subgroups	Number of students
Reactions	Positive	general comments	15
		usefulness	6
		interaction	3
		teaching style	2
	Negative	hours needed	1
		mathematical artifices	1
	Neutral	topic	3
		timing	3
		other aspects	3

Usefulness of the applications:

"Now I find that mathematics can be useful."

"I thought the course was very useful and dynamic, and I think it will have very useful applications in coming years."

"Very useful examples for future years of undergraduate study."

Interaction with other subjects in chosen careers paths:

"The problems were motivational because you can see the usefulness of mathematics in daily life, and they clearly show the interaction that exists with other subjects."

"I have taken courses in which the applications dealt with here were relevant."

Teaching style:

"Very directly applicable to my undergraduate professional career; it renewed a taste for mathematics and it was well taught, guiding the solutions to the exercises and not working them all out."

"If these topics were omitted, the course would just be another standard math course, a 'hard' subject filled only with methods, calculations and numbers."

Only 2 students had negative opinions, one concerned the total amount of hours needed for understanding these topics, while the other criticized the mathematical intricacies involved in solutions:

> "Classes are well taught but in my opinion the amount of hours is not enough."
>
> "In the applications problems, mathematical artifices are used by teachers but I can't cope with them."

The rest of the responses are considered as "neutral reactions", since they showed both positive and negative feedback about the course. A first sub-group had a positive opinion about the course but added about the application problems, as in the following example:

> "The course is O.K., but I would add chemical examples and put less physical ones."

A second sub-group agreed with the idea of spending time in applications but they suggest modifying the timing:

> "I agree with the idea of teaching applications but it would be better to dedicate less time in the first part of the course and more during the second one."

The last sub-group, made comments about other aspects not strictly related with the course itself. The following is an example of this kind of responses:

> "Applications should be taught in Mathematics III as well as in other subjects of the career where more details of the problems could be explained."

Results Obtained in Teacher Training Courses

In the last course-workshop, held in Colima, the students were asked about their views of the course, the inverse modeling problems, the pitch and pace, course organization, etc. The following interesting comments were received:

Student 1: The workshop was interesting because we discovered how we can create a new problem from an existing one, in a reverse way. Besides, we learnt that the problems may not have solution or can have diverse solutions and they can be solved through different methods.

Student 2: I enjoyed participating in this workshop because I learnt new things. We learnt how to think about the (direct) problem first, and to pose the related inverse problem afterward.

Student 3: The problems are very interesting since they are related to real-life.

Student 4: This workshop was very interesting and exciting and I like the mode we used to go from one direct problem to an inverse one, in that way we recognized daily life situations where we can apply that knowledge.

The students reacted very positively not just to the course they were taught, but in particular, they appreciated all aspects of the modeling and application problems whether, direct or inverse.

CONCLUSION

This last section includes the conclusions obtained from our experiences in both undergraduate engineering courses and in teachers and prospective teachers training courses. The first part will be devoted to our experiences with inverse modeling problems in undergraduate engineering courses, first in Uruguay and later in Argentina.

It can be concluded from the previous section, that the students' feedback was positive to the whole ambiance of the course. The comments suggest that the interest in the course is due to the applied approach, as seen in the first, third and fourth response in the previous section.

The students' reactions mentioned previously cannot be considered surprising. In fact, in engineering degree courses – and no doubt in several others – there is an invisible dividing line between "basic" and "technological" subjects. Indeed, in informal conversations with students (and former students), they commonly talk about their "urgency" to reach as quickly as possible the technical subjects at the end of their degree course. They associate several positive characteristics with these subjects: their evident usefulness, their applicability – in many cases, in the short-term – and many of them also take a favorable view of the minimum amount of time dedicated to proofs and first principles in these subjects.

At the opposite end of the spectrum are the "basic" subjects, particularly the different branches of mathematics like Calculus, Linear Algebra and perhaps in a different category, the Differential Equations courses. These subjects are where students often find the other side of the coin: rigorous theoretical deduction, coherent and frequently elegant developments of mathematical arguments, strictly consistent notation, etc. At the same time, they also find some less desirable characteristics, such as: dubious applicability, learning concepts that are under-utilized, a kind of “lack of fit” between a given concept and its practical application, and in general a certain disconnection with technical subjects.

Thus, the big challenge appears to be the connection between the expository rigor of basic sciences – especially mathematics – and the practical applicability of technological subjects. Our experiences with engineering students in Uruguay and more recently in Argentina, suggest that inverse modeling problems seem to be useful in order to create a connection between practice and theory.

For instance, working with the pirate problem, students first tried to take different points to place the flag on the island, then they followed the pirate’s procedure, and discovered that they always obtained geometrically the same point. After that, they performed the corresponding algebraic calculations in order to confirm these results. Finally, in the oral presentation of their results, they were asked to give formal proofs. It is important to remark that all the students groups succeeded in the geometrical visualization, which corresponds to the “embodiment” world of Mathematics (Tall, D., 2006) and most of these groups also succeeded in confirming their results performing the algebraic calculations, i.e., a procedural approach, corresponding to the “proceptual world” of Mathematics (Tall, D., 2006). For these first year students groups, it was not so easy to arrive at a formal proof and they needed some encouragement from their teachers in order to complete this task in their oral presentation.

As David Tall observed (Tall, D., 2006), “…in the embodied world a warrant for truth occurs first through seeing something is true…” while “…in the proceptual world, a warrant for truth arises through calculating that a result is correct…”. Our experiences with the students working on the pirate’s problem show that almost all of them can get these warrants for truth with a minimum help from their instructors. The situation changes when the students are asked to construct and prove results formally. At least, for first year university students this path into the “formal world” of Mathematics needs a huge range of experience and techniques which they are still developing.

Another important issue about the pirate’s problem, that deserves to be mentioned is that it gives geometrical meaning to complex numbers and several operations like multiplying by i or $-i$. According to David Tall (Tall, D., 2004) "…the case of complex numbers which, at their inception, had no conceptually embodied meaning. (…) Gauss in his 1799 doctoral thesis and Argand in 1806 re-invented complex numbers as points in the plane…" The pirate problem gives a good opportunity to show these connections to the learners in a motivating context and can be considered an important outcome for the students.

It is important to note that several of these problems – and others not mentioned here in this chapter – gave the students the opportunity to establish connections between the three worlds of Mathematics (embodied, proceptual and formal) and also, they helped to create relations between theory and practice.

The other part of our work with inverse modeling problems was the work with teachers and prospective teachers. In those training courses, it was not difficult for the teachers and prospective teachers to transform a direct modeling problem into an inverse one. This process enriches the problem and allows them to explore existence, uniqueness and stability which are not so interesting in direct problems (moreover, often these properties are trivial in direct problems). As a consequence, a greater understanding of the problem is needed.

The engineering students' reactions in Uruguay and Argentina (Martinez-Luaces, V., 2009a, 2013) and those obtained with the prospective teachers in Colima, Mexico can be considered as rather similar since in all cases students appreciated that these problems helped them to obtain a deeper understanding, analyze other points of view, create relations between theory and practice and make connections with real-life situations.

Further research will consist in preparing other inverse modeling problems, to be 'tested' with students and prospective teachers in other courses and/or universities in different countries. In Latin America regarding the usual difficulties about the lack of material and human resources, we tried to include real-life problems in our courses and make more strong connections between theory and practice. In order to achieve this goal, inverse problems and mathematical modeling nowadays have and will continue playing an important role.

ACKNOWLEDGMENTS

The author wishes to thank Marjorie Chaves and Anne D´Arcy-Warmington for their valuable contribution to this work.

REFERENCES

Blum, W., (2002). ICMI Study 14: applications and modelling in mathematics education – discussion document. *Educational Studies in Mathematics, 51 (1-2),* 149-171.

Bunge, M. (2006). *Problemas directos e inversos*, Retrieved in March, 2016 from the web page: http://grupobunge.wordpress.com/2006/07/20/119

Corey, E.J., (1991). The logic of chemical synthesis: Multistep synthesis of complex carbogenic molecules (Nobel Lecture). *Angewandte Chemie International Edition in English, Vol. 30*, Iss. 5, 455-465.

Courant, R. (1937). *Differential and Integral Calculus.* Volume I, 2nd Edition. London and Glasgow, UK:. Blackie and Son Limited.

Gomez, A. and Martinez-Luaces, V., (2002), Evaluacion docente utilizando analisis multivariado. *Acta Latinoamericana de Matematica Educativa, 15,* 1016–1021.

Gonzalez, B.E., (2001), *La preparacion del profesor para la utilizacion de la modelacion matematica en el proceso de enseñanza-aprendizaje.* (Tesis para la obtencion del grado de Doctor en Ciencias Pedagogicas). Santa Clara, Cuba: UCLV.

Groetsch, C.W. (1999). *Inverse problems: activities for undergraduates,* Washington, D.C.: Mathematical Association of America.

Groestch, C.W. (2001). Inverse problems: the other two-thirds of the story, *Quaestiones Mathematicae, Supplementary Issue (1)*, 89-93.

Guerasimov, Y.A. et al. (1995), *Physical Chemistry, 2nd Ed.*, Boston, USA: Houghton-Mifflin.

Guzman, M., 1991, *Para pensar mejor.* Barcelona, España: Labor-MEC.

Guzman Ozamiz, M., (2010). *El rincón de la pizarra. 2nd. Edition.* Spain: Ed. Pirámide.

Hernandez, M.R., (1997), Niveles epistemologicos en el aprendizaje de las matematicas. *Educacion Matematica, 9*, 43–52.

Mason, J., Burton, L. and Stacey, K., (1988), *Pensar matematicamente* (Barcelona, España: Labor-MEC).

Martinez-Luaces, V., (1998), Matematica como asignatura de servicio: algunas conclusiones basadas en una evaluacion docente. *Numeros. Revista Española de Didactica de Matematica, 36*, 65–74.

Martinez-Luaces, V., (2005), Engaging Secondary School and University Teachers in Modelling: Some Experiences in South American Countries, *International Journal of Mathematical Education in Science and Technology, Vol. 36 Issue 2-3*, pp. 193-205.

Martinez-Luaces, V., (2007). Inverse-modelling problems in chemical engineering courses, in *Vision and change for a new century. Proceedings of the Sixth Southern Hemisphere Conference on Undergraduate Mathematics and Statistics Teaching and Learning*. Montevideo, Uruguay: ISC-Delta.

Martinez-Luaces, V., (2009a). Modelling and inverse-modelling: experiences with O.D.E. linear systems in engineering courses. *International Journal of Mathematical Education in Science and Technology 40(2)*, 259-268.

Martinez-Luaces, V., (2009b). Modelling, applications and inverse modelling: innovations in differential equations courses. In D. Wessels y C. Snyman (Eds.), *Proceedings of Southern Right Delta '09, Seventh Southern Hemisphere Conference on Undergraduate Mathematics and Statistics Teaching and Learning*. South Africa: ISC-Delta.

Martinez-Luaces, V., (2009c). Modelling and inverse modelling: innovations with second order P.D.E. in engineering courses. In D. Wessels y C. Snyman (Eds.), *Proceedings of Southern Right Delta '09, Seventh Southern Hemisphere Conference on Undergraduate Mathematics and Statistics Teaching and Learning*. South Africa: ISC-Delta.

Martinez-Luaces, V., (2011a). Cursos de Cálculo y pre-Cálculo: el papel de los problemas inversos, las aplicaciones y el modelado matemático. *Reunión de Educación Matemática (REM XXXIV).* Tucuman, Argentina.

Martinez-Luaces, V., (2011b). Problemas inversos: los casi olvidados de la Matemática Educativa. *Acta Latinoamericana de Matematica Educativa, 24*, 439-447.

Martinez-Luaces, V., (2011c). Mathematics and its connections with technical subjects, the real world and professional life. *Proceedings of Volcanic Delta 2011*, pp. 238-246.

Martinez-Luaces, V., (2012a). Chemical Kinetics and Inverse Modelling Problems, in *Chemical Kinetics*, Chapter 3, pp. 61-78. Rijeka, Croatia: In Tech Open Science.

Martinez-Luaces, V., (2012b). *Problemas inversos y de modelado inverso en Matemática Educativa*, Saarbrücken, Germany: Editorial Académica Española.

Martinez-Luaces, V., (2013). Inverse modeling problems in Linear Algebra undergraduate courses. *International Journal of Mathematical Education in Science and Technology, Vol.44, Iss. 7*, 1056-1064.

Martinez-Luaces, V., (2015). First Order Chemical Kinetics Matrices and Stability of O.D.E. Systems in *Advances in Linear Algebra Research.* Chapter 10, pp. 325-343., New York, U.S.A.: Nova Publishers.

Martinez-Luaces, V. (2016), Stability of O.D.E. systems associated to first order chemical kinetics mechanisms with and without final products, *Konuralp Journal of Mathematics. Proceedings of the 4th International Eurasian Conference on Mathematical Sciences and Applications*, in press.

Martinez-Luaces, V., Camarena, P. and Sallet, M., (2004), Modelos matematicos, *Acta Latinoamericana de Matematica Educativa, 17(1),* 423–427.

Martínez-Luaces, V. and Guineo Cobs, G., (2002). Un problema de Electroquímica y su Modelación Matemática, *Anuario Latinoamericano de Educación Química. Vol XVI*, pp. 272 – 276.

Oliveira e Silva, T., Herzog, S. and Pardi, S., (2014). Empirical verification of the even Goldbach conjecture and computation of prime gaps up to $4 \cdot 10^{18}$. *Mathematics of Computation 83.288*, 2033-2060.

Schoenfeld, A.H. (1994), *Ideas y tendencias en la resolucion de problemas.* Buenos Aires, Argentina: Olimpiada Matematica Argentina.

Tall, D.O. (2004). The three worlds of Mathematics. *For the learning of Mathematics, 23 (3),* 29-33.

Tall, D.O. (2006). A theory of mathematical growth through embodiment, symbolism and proof. *Annales de Didactique et Sciences Cognitives,* IREM de Strasbourg, 11, 195-215.

Varga, R.S. (2004), *Geršgorin and His Circles.* Berlin, Germany: Springer-Verlag.

Vollhardt, K. Peter C. and Schore, N.E., (2007). *Organic chemistry: structure and function*. Macmillan.

Wilhelmy, L. (1850), Über das Gesetz, nachwelchem die Einwirkung der Säuren auf den Rohrzuckerstattfindet, *Pogg. Ann. Vol. 81*, pp. 413-433.

Zambelli, S., (2012), Chemical Kinetics, a Historical Introduction, in *Chemical Kinetics*, Rijeka, Croatia: InTech Open Science Eds.

Zinola, F., Méndez, E. and Martinez-Luaces, V., (1997). Modificación de estados adsorbidos de Anhídrido Carbónico reducido por labilización electroquímica en superficies facetadas de platino. *Proceedings of X Congreso Argentino de Fisicoquímica*, Tucumán, Argentina.

BIOGRAPHICAL SKETCH

Name: Victor Martinez-Luaces

Affiliation: J. Ricaldoni Foundation, Faculty of Engineering, UdelaR, Montevideo, Uruguay

Education:
Degree in Chemistry, Faculty of Chemistry, UdelaR, 1982
Chemical Engineer, Faculty of Engineering, UdelaR, 1985
Degree in Mathematics, Faculty of Sciences, UdelaR, 1992
Post-graduate studies in Mathematics at IMPA (Brazil)
Post-graduate studies in Mathematics Education at UGR (Spain)
Mail address: victorml@fing.edu.uy

Research and Professional Experience:
First book chapter, 1982, Uruguay
First research paper, 1988, Journal of Chemical Education, U.S.A.
Since then, several papers have been published worldwide in the field of Mathematics Education, Chemistry, Engineering and Mathematics in journals, conference proceedings and books.
More than 10 books, 50 journal papers and 80 proceedings papers.

Professional Appointments:
Faculty of Chemistry, UdelaR Head of the Mathematics Department 1996 - 2002
Faculty of Engineering, UdelaR 1988 - 1997
Faculty of Economics, UdelaR 1990 - 2006
Faculty of Sciences, UdelaR 1999- 2000
CIEP (Education and Research Institute) Rio Gallegos, Argentina 2006-2008
Since 2009 Mathematical and Engineering Consultant at J. Ricaldoni Foundation, UdelaR.

Honors:

- Journal of Chemical Education – (Ethyl Corporation, 1988).
- Best teacher award (Faculty of Engineering, 1992).
- Best teacher award (Faculty of Engineering, 1996).
- 2nd prize RELME 13 (Santo Domingo, Dominican Republic, 1999)
- 2nd prize RELME 14 (Panama, 2000)
- 2nd prize RELME 15 (Buenos Aires, Argentina, 2001)
- 1st prize RELME 16 (La Habana, Cuba, 2002)
- 1st prize RELME 19 (Montevideo, Uruguay, 2005)
- Distinguished visitor of the first city in the Americas (Santo Domingo, 1999)
- Co-chair of the TSG 13 at ICME-12 in Seoul, Korea 2012
- Co-chairs of the TSG 16 at ICME-13 in Hamburg, Germany 2016.
- Reviewer for several journals as ZDM, iJMEST, IJRUME, ALME (Mexico), Union (Spain) and Numeros (Spain), American Journal of Applied Mathematics and Statistics and American Journal of Educational Research, among others.
- Part of the Editorial Board for journals like IJRUME, Numeros, Union and Global Education Review.

Publications Last 3 Years:

A) Books

1) Martinez-Luaces, V., (2015). First Order Chemical Kinetics Matrices and Stability of O.D.E. Systems. In *Advances in Linear Algebra Research*. Chapter 10, pp. 325-343. Nova Publishers, New York, U.S.A.
2) Martinez-Luaces, V., (2012). *Problemas inversos y de modelado inverso en Matemática Educativa*, Editorial Académica Española, Saarbrücken, Alemania.
3) Martinez-Luaces, V., (2012). Chemical Kinetics and Inverse Modelling Problems, in *Chemical Kinetics*, Dr Vivek Patel (Ed.), In Tech Open Science.

B) Journal publications

1) Martinez-Luaces, V. (2016), Stability of O.D.E. systems associated to first order chemical kinetics mechanisms with and without final products, *Konuralp Journal of Mathematics*, in press.
2) Martinez-Luaces, V. (2015), Stability of O.D.E. solutions corresponding to chemical mechanisms based-on unimolecular first order reactions, *Mathematical Sciences and Applications E-notes. Vol. 3, Nº 2*, pp. 58-63.
3) Ohanian, M. and Martinez-Luaces, V., (2014). Corrosion potential profile simulation in a tube under cathodic protection. *International Journal of Corrosion*. http://dx.doi.org/10.1155/2014/102363
4) Martinez-Luaces, V., (2013). Inverse modelling problems in linear algebra undergraduate courses. *International Journal of Mathematical Education in Science and Technology, 44, 7*. Pp 1056-1064.
5) Martinez-Luaces, V. and Oates, G., (2013). Calculus from around the world. *International Journal of Mathematical Education in Science and Technology, 44, 5*. pp 621 – 623.
6) Ohanian, M. and Martinez-Luaces, V., (2012). Distribución de potencial electroquímico en corrosión interna de tuberías: presentación de modelo en condición de flujo y resolución por Series de Fourier-Bessel. *Revista del Instituto de Matemática. XVI. Año 8*. Pp. 75-82.
7) Martinez-Luaces, V. and Velázquez, B., (2012). A course on experimental design for Biotechnology students. *Community for undergraduate learning in the Mathematical Sciences, No. 5*. pp 40 – 47. www.math.auckland.ac.nz/CULMS
8) Ospitaletche-Borgmann, E. and Martínez Luaces,V., (2012). La Matemática como idioma y su importancia en la enseñanza y aprendizaje del Cálculo. *Números. Revista Española de Didáctica de Matemática. 79*. Pp. 7 – 16.

C) Proceedings and other publications

1) Martinez-Luaces, V., (2016). Mathematical models for Chemistry and Biochemistry service courses. Proceedings of CERME 2015. Prague, Czech Republic, in press.

2) Martinez-Luaces, V. and Noh, S., (2015). Report of the Topic Study Group 13 about Teaching and Learning of Calculus, *Intellectual and attitudinal challenges, Proceedings of ICME-12*. Korea, pp. 447-452.
3) Martinez-Luaces, V., (2013). Problemas inversos y de modelado inverso: posibilidades y potencialidad en Matemática Educativa. *Memoria del Congreso internacional en Ciencias de la Educación*, Universidad de Colima, México, pp 15-23.
4) Martinez-Luaces,V. and Ohanian, M., (2012). Resolución analítica de distribución de potencial electroquímico en geometría cilíndrica con aplicación a corrosión interna de tuberías. *Actas de EMCI-XVII*, UTN, Buenos Aires, Argentina.
5) Martinez-Luaces,V., (2012). Inverse modelling problems in Calculus courses. *ICMIXII, TSG-13*, Seoul, Korea. http://www.icme12org/upload/Upfile2/TSG/0386.pdf6

In: Teaching and Learning
Editor: Malcolm Vargas
ISBN: 978-1-63485-228-9

Chapter 8

FORMATIVE FEEDBACK RESEARCH: REVIEW OF STUDIES IN ONLINE AND BLENDED LEARNING ENVIRONMENTS

Anna Espasa[1,*], Maria José Rochera[2] and Teresa Guasch[1]
[1]Open University of Catalonia, Spain
[2]University of Barcelona, Spain

ABSTRACT

Research studies on feedback and technology in learning evidence that technology can play an important role in facilitating feedback; however, it is still not clear how this is done nor is there agreement on what the most pressing issues for research in this area are. In order to address this gap, this paper aims to analyse the research on formative feedback in online or blended environments carried out between 2008 and 2012. A threefold analysis was performed, corresponding to three questions: what theoretical assumptions have these research studies adopted? What objectives and problematics were addressed? And lastly, what methodologies were used? The answers to these questions will provide a general overview of research in this area as well as identify trends that may guide our decision-making in terms of setting priorities

*Corresponding author: Anna Espasa. Open University of Catalonia, Rambla del Poblenou, 156, 08018, Barcelona (Spain) Email: aespasa@uo.edu.

for future research and improving feedback practices in environments requiring the use of technology.

Keywords: formative feedback, feedback, personalised feedback, dialogic feedback, online learning environments, higher education

INTRODUCTION

To understand the main challenges facing e-assessment in the 21st century and to illustrate how technology-enabled assessment is evolving, Redecker and Johannessen refer to a model that had been used in previous research to forecast four successive stages or generations of computer-based assessment (Redecker and Johannessen, 2013). While the first two generations had examined how computers would be used to improve the efficiency of testing procedures, the third and fourth generations highlighted the importance of embedding assessment in teaching and learning. Specially, in Fourth Generation, learning systems would be able to provide personalized feedback. Redecker and Johannessen propose that "assessment strategies need to be better harmonised with 21st century approaches by re-focussing on the importance of providing timely and meaningful feedback to both learners and teachers" (pp. 80).

This new approach proposed by Redecker and Johannessen (2013) is encompassed in the dialogic perspective of feedback (Carless et al. 2011, Nicol, 2010). According to Carless (2015), dialogic feedback is defined "as a process rather than a product; it emphasises dialogue which could be with a teacher, peer, other contact, or the self through some inner dialogue; and emphasis is placed on what students do with feedback in relation to their current of future work and/or modifying the kinds of learning strategies they are adopting" (pp192). Yang and Carless (2012) define three dimensions of feedback to enhance learning: cognitive, social-affective and structural. Cognitive dimension refers to the type of information that feedback provides and is related to the learning content; social-affective dimension is defined as dialogic interaction within a trusting atmosphere to promote self-regulation of learning and the structural dimension includes timing and forms of feedback presentation (i.e., written feedback, audio-feedback…). Narciss (2008) proposes similar feedback components: content, structure but she also highlights feedback function, which is understood as the aim with which feedback is given. In relation to this, it is worth noting that current theoretical approaches tend to conceptualise feedback from a multidimensional perspective and they take into account various aspects of feedback (Coll, Rochera and De Gispert, 2014; Espasa, Guasch and Alvarez, 2013).

As Hattie and Gan observe, the functions feedback can serve depend on the particular learning perspective being used (Hattie and Gan, 2011). Reflections on perspectives on learning and the nature of feedback presented by these authors are very useful for the purposes of our review. They analyse four major psychological perspectives describing learning and its relation to feedback: objectivism, information processing, socioculturalism and visible learning and teaching. The objectivist perspective of learning considers that external feedback provides corrective information; the information processing perspective uses feedback to help learners to "build internal understanding through connections with [learners'] prior experiences, mental structures and beliefs"; the sociocultural perspective of learning understands feedback as "a reciprocal and dialogical process of co-construction of meaning" —this is precisely the approach dialogic feedback is grounded in—; and visibility learning and teaching holds that feedback is more powerful when it makes student learning visible to both teachers and students and it considers that the feedback quality "depends on the changes to teaching and learning strategies of the teacher, learner, and peers to effect the learning goals" (pp 251). However, despite the fact that these reviews examine several key issues on feedback research, they contain no systematic evidence on the main theoretical and methodological approaches that support the use of formative feedback in online and blended environments.

In the specific context of online environments, feedback is considered one of the most important supports that students need for promoting learning (Lee et al., 2011). A recent review of literature on the wide-reaching topic of online formative assessment in higher education concludes that "effective formative feedback is critical to the effectiveness of online formative assessment particularly in relation to aspects of immediacy, adequacy, and interactivity" (Gikandi, Morrow and Davis, 2011, p. 2447). The same review also proposes that online settings enhance opportunities for immediate and ongoing formative feedback in ways that face-to-face settings cannot because computer application software provides detailed written feedback and offers opportunities for feedback visibility to all those involved in the process (in online asynchronous discourse, for example, writers can revisit their own contributions and those of others). Moreover, Hepplestone et al. (2011) has directly addressed the question of how technology supports student engagement with feedback and has identified its presence in a number of areas, including the use of computer-assisted assessment with instant feedback, of peer feedback supported by electronic tools, and of audio feedback to provide more complete and detailed feedback than written formats can provide.

These studies evidence that technology may contribute positively to the enhancement of feedback; however, it is still not clear how this is done nor is there agreement on what the most pressing issues for research in this area are. In order to address this gap, this paper aims to analyse the research on formative feedback in online or blended environments carried out between 2008 and 2012. A threefold analysis was performed, corresponding to three questions: what theoretical assumptions (i.e., theoretical approaches, feedback definition and the use of ICT to give feedback) have these research studies adopted ? What objectives and problematics were addressed? And lastly, what methodologies were used? The answers to these questions will provide a general overview of research in this area as well as identify trends that may guide our decision-making in terms of setting priorities for future research and improving feedback practices in environments requiring the use of technology.

METHODS

A review article has been variously defined as a text that synthesizes and critically evaluates the research conducted in a specific area (Booth, Papaioannou and Sutton, 2012). As such, review articles allow reviewers to account for the state of understanding on a topic in a given period.

We reviewed the topic of formative feedback in online and blended learning environments in two main stages, first by searching for and selecting a corpus of papers and second by analysing that corpus according to the quality assessment categories and criteria proposed by Boote and Beile (2005).

Searching for and Selecting the Literature

The purpose of this stage was to search for and select the papers to be included in the study. This was particularly important because the decisions taken here would determine what would or would not be available for analysis in the final sample (Boote and Beile, 2005). The following steps were required: first, the definition of a set of criteria including keywords, the time period, the area or field, the type of production and the search field; second, the definition of inclusion and exclusion criteria for the papers that would be found (Boote and Beile, 2005; Hart, 1999). Authors such as Bruce propose a total of eight principles for judging the adequacy and quality of selected literature: topicality, comprehensiveness, breadth, exclusion, relevance, currency, availability, and authority (Bruce, 2001).

Different search terms were used relating to feedback and ICT in education, including the descriptors *formative feedback*, *teacher*, *students*, *peer*, *learning*, *online*, *blended*, *ICT* and *technology*. To delimit the search, these descriptors were combined in successive cycles (for example, *formative feedback* and *teacher* and *online* followed by *formative feedback* and *teacher* and *online*, in turn followed by *formative feedback* and *teacher* and *ICT*, and so on). Two authoritative electronic databases were consulted, the Education Resources Information Center (ERIC) and the world's largest abstract and citation database of peer-reviewed literature, Scopus. The search focused on the field of social sciences and the humanities in higher education in the time period 2008–2012 and to ensure the quality of the review, only peer-reviewed sources were considered. These search criteria yielded a total of 102 papers and the search process was continued until the results revealed no significant new papers.

Having concluded the search, the selection of a corpus from these 102 papers was completed by the authors of this article selecting the papers taking into account two main criteria: (1) papers addressing the educational importance of feedback rather than its technical features and (2) papers addressing this importance empirically. We then screened the result in three successive phases (Booth, Papaioannou and Sutton, 2012): first, we included or excluded papers according to the relevance of their titles; next, and where relevance could not be judged by title alone, we included or excluded papers according to the relevance of their abstracts; and finally, where relevance could still not be judged by either title or abstract, we examined the full text of those papers whose abstracts appeared to conform to our inclusion criteria. The authors of this article, working independently, achieved an agreement of 90% in the papers' selection. After a negotiating process, they finally agreed 100% on the final article sample. At the end of this selection process, we had identified a total of 24 papers that would constitute our final sample and be used for analysis (Table 1).

Table 1. Summary of the 24 empirical papers included in this review. This numbering will be used to identify the papers reviewed

Authors and year	Context of the study	Discipline
1. Alvarez, I. Espasa, A. and Guasch, T. (2012).	Online	E-learning
2. Berlanga, A.J., Kalz, M., Stoyanov, S., van Rosmalen, P., Smithies, A. and Braidman, I. (2011)	Online	Medical degree/ Psuchology degree
3. Crook, A., Mauchline, A., Maw, S., Lawson, C., Drinkwater, R., Lundqvist, K. and Orsmond, P. (2012)	Online	Arts, Humanities, Bussiness, Law, Life Science and Science
4. Dekhinet, R. (2008).	Online	English as a Second Language (ESL)
5. Dexter, S. (2010).	Online	Education/Educational Technology
6. Espasa, A. and Meneses, J. (2010).	Online	Engineering/Law/Psychology and Education/ Economics/ Communication
7. Gallien, T. and Oomen-Early, J. (2008).	Online	Health education
8. Getzlaf, B., Perry, B., Toffner, G., Lamarche, K and Edwaeds, M. (2009).	Online	Nursing
9. Hatziapostolou, T and Paraskakis, I. (2010).	Online	Computer Science
10. Hwang, A. and Arbaugh, J.B. (2009)	Online	Business
11. Jones, N., Georghiades, P., Gunson, J. (2012).	Blended	Economics
12. Lin, W.-C., Yang, S.C. (2011).	Online	English writing courses.
13. Dysthe, O. and Lillejord, S. (2012).	Online	Education
14. Macgregor, G. Spiers, A. Taylor, Ch, (2011).	Online	Business Management and Information degree course.
15. McVey, M. (2008).	Online	Child and Adolescent Development
16. Moradi, M. R. and Karimpour, Z. (2012).	Online	English as a Foreign Language (EFL)
17. Mostyn, A., Meade, O., Lymn, J.S. (2012)	Blended	Pharmacology
18. Oskoz, A. (2009).	Online	Spanish as a second language.
19. Parkin, H.J., Hepplestone, S., Holden, G., Irwin, B. and Thorpe, L.(2012).	Online	Computer networks, psychology, diagnostic radiography and events management.

Authors and year	Context of the study	Discipline
20. Shroff, R.H. and Deneen, Ch. (2011).	Online	Information Systems
21. Van der Pol, J., Van den Berg, I., Admiraal, W. and Simons, R.J. (2008).	Online	Health Care Education
22 Wieling, M. B. and Hofman, W. H. A. (2010).	Blended	Law
23. Wood, K. A., Moskovitz, C. and Valiga, T. M. (2011).	Online	Nursing
24. Xu, Y. (2010).	Blended	Special education

Analysing the Literature

The purpose of this stage was to summarize the literature that had been chosen. The 24 papers were qualitatively analysed using thematic analysis so that they could be assigned categories and contrasted dialectically (Booth, Papaioannou and Sutton, 2012). Three thematic blocks were established: a first thematic block which included the theoretical perspectives, the conceptualization of formative feedback and the use of online tools to give and receive formative feedback (Table 2); a second thematic block which included the main aims of formative feedback in online environments, the problems addressed and the findings related to these aims (Table 3); and a third thematic block which included methods for analysing formative feedback in online environments (Table 4). To carry out the content analysis, an open coding process per each thematic block was used. Categories for the analysis were identified inductively. They were obtained from the explicit information in the papers. After an initial outline of categories had been constructed, a dialectical and continuous process was used to reformulate the categories in contrast with the data and establish definitive categories (Tables 2, 3 and 4). Finally and in order to meet the reliability and consistency criteria required for the content analysis, an inter-judge procedure was used and the two authors coded the 24 papers by category obtaining in all cases a 100% agreement.

Once this initial synthesis had been performed and in order to draw conclusions, the analysis of the literature in terms of what had been found and what was missing (Booth, Papaioannou and Sutton, 2012) was carried out. Thus, we were able to summarize the main results obtained in previous research on the topic and also draw attention to areas of interest for future research.

Table 2. Theoretical assumptions of formative feedback in online and blended environments across the 24 empirical studies included in this review

Theoretical perspectives and conceptualization of formative feedback			
Theoretical perspectives explicitly cited	**Number of the papers (*see Table 1 to identify the papers*)**	**Frequency of papers (N = 24)**	**% of papers**
Formative assessment, Formative feedback	1, 2, 3, 5, 6, 7, 8, 9, 11, 14, 15, 19, 21, 23, 24	14	58.3%
Sociocultural perspective	1, 4, 6, 12, 13, 18	6	25%
Online learning communities	8, 13	2	8.3%
Feedback-seeking theory	10	1	4.2%
No explicit	16, 17, 20, 22,	3	16.7%
Feedback conceptualization (main focus)			
Focus on content	1, 3, 6, 7, 8, 9, 13, 15, 21	9	37.5%
Focus on time	2, 3, 7, 8, 15	5	20.8%
Focus on aim/Function	1, 5, 6, 10, 20, 23	6	25%
No explicit	2, 4, 11, 12, 14, 16, 17, 18, 19, 22	10	41.67%
Digital tools to formative feedback			
Using technology to produce written feedback	1, 4, 5, 6, 7, 8, 9, 12, 13, 15 16, 18, 19, 20 21, 24	15	62.5%
Using technology to produce video feedback (continuous feedback)	3, 11, 22	3	12.5%
Using technology to produce audio feedback	10, 14, 23	3	12.5%
Digital tools to formative feedback			
Using technology to obtain evidence on learning as a basis for providing feedback	2, 17	2	8.3%

The categories in each question are not exclusive; that is, one paper can include different ideas about theoretical perspective and can simultaneously introduce two o more different characteristics on feedback. For this reason, in each question, the frequencies of each category do not total 24; and the percentages do not total 100.

RESULTS

The results are going to be presented according to the three thematic blocks proposed in the previous section: (1) theoretical assumptions about formative feedback in online environments; (2) aims or problems analysed on formative feedback research in online environments and (3) methodological approaches used.

Theoretical Assumptions about Formative Feedback in Online and Blended Environments

With regards to the *theoretical assumptions*, three issues have been analysed: the theoretical approach, the conceptualization of feedback and the use of technology to produce formative feedback.

Through the analysis of the 24 selected papers, different *theoretical perspectives* were identified. As shown in Table 2, more than 50% of papers looked at the notion of formative assessment or formative feedback as a key element to promote learning (58.3%) in online or in blended environments. For example, Berlanga et al. (2011) [2] point out that feedback is a key element in formative assessment and it “can be defined as information provided by an agent (e.g., teacher, peer, book, parent, tool) regarding aspects of performance or understanding (...) with the aim of modifying thinking in order to improve learning”. (pp. 12).

In 25% of the papers analysed feedback was linked to a sociocultural perspective. This term is used in a broad way. For example, in Alvarez, Espasa and Guasch (2012) [1], formative feedback is related to the teacher’s support to knowledge construction. Another example is the paper by Lin and Yang (2011) [12], where peer feedback is linked to peer assistance and mutual scaffolding in collaborative dialogues. Dysthe and Lillejord (2012) [13] refer to a dialogical view of learning and technology. Among the less frequent theories used were the Online Learning Communities models (8.3%). For example, Getzlaf et al. (2009) [8] refer to the Community of Inquiry framework of the authors Garrison, Anderson and Archer, and Hwang and Arbaugh (2009) [10] uses Feedback-seeking theory. Finally, 16.7% of the papers do not mention the theoretical approach adopted for their study, which makes it difficult to understand in greater detail the significance of their results.

Concerning the *conceptualization of feedback*, 37.5% of papers define feedback focusing on content. In the majority of papers, formative feedback refers to any information communicated to the learner to help him or her to improve and enhance performance. Some papers include different types of information with respect to the content of feedback. For example, feedback may comprise corrective information and suggestions (Alvarez, Espasa and Guasch, 2012 [1]); information about errors, correct answers, guidelines to improve learning/additional resources (Espasa and Meneses, 2010) [6]); information with respect to different aspects of a task, enabling individuals to see areas of inadequate performance for the next round of improvement (Hwang and Arbaugh, 2009 [10]); explicit information on how performance can be improved (Crook et al., 2012 [3]; McVey, 2008 [15]); clarification of the problem and explanation of concepts and structure or style (Van der Pol et al., 2008 [21]); and lastly, feedback might also include product oriented and process oriented information (Getzlaf et al., 2009 [8]). There are papers which suggest that feedback might be tailored (Gallien and Oomen-Early, 2008 [7]), is provided as a result of a learning-oriented action (Getzlaf et al., 2009 [8]; Hatziapostolou and Paraskakis, 2010 [9]) and might be productive, that is to say, to be able to engage participants in a dialogical process with others –peers or teachers- as pointed out by Dysthe and Lillejord (2012) [13].

Several papers, 25% to be precise, highlight the different functions of feedback: cognitive, metacognitive, regulative and motivational. For example, feedback can help students engage in more sophisticated synthesis of sources and ideas (Wood, Moskovitz and Valiga, 2011 [23]); it increases student satisfaction and motivation (Shroff and Deneen, 2011 [20]) and promotes metacognition (Dexter, 2010 [5]); or the regulation of learning (Hwang and Arbaugh, 2009 [10]).

In terms of the timing, 20.8% has argued that feedback might be offered during the teaching and learning process (Espasa and Meneses (2010) [6], Getzlaf et al., 2009 [8]); immediately after its referent (McVey, 2008 [15]) and it can be provided in a manner which the students feel it shouldn't be too late to be useful (Crook et al. (2012) [3]). In a few papers, timing is considered from the instructors' point of view, in the sense that producing feedback is regarded as very time-consuming, and the use of technology is seen as an effective means to reduce the time taken to elaborate feedback. For example, Crook et al. (2012) [3] indicate that "the provision of feedback can be a very repetitive process and often very time consuming especially where class sizes are large" (pp 386) and that "by providing many of the key messages to all students in one generic video, it has potential to make the process of producing individual feedback more efficient as the main points have already been covered, thus allowing more time for staff to spend on delivering tailored feedback to individual students" (pp. 394).

With respect to the *use of online tools to deliver formative feedback*, 62.5% of papers propose different online tools to produce written feedback in a online learning environment. 12% of papers report the use of online audio for feedback purposes, and coincidentally, the same percentage of papers (12%) report the use of online video for the same purpose. For example, Wood et al. (2011) [23], use Wimba Voice Recorder via Blackboard educational software for the transmission of feedback.

A few papers looked at using technology to obtain evidence of learning as a basis for providing feedback (8.3%). For example, to use language technologies to create concept maps automatically from texts in online environments (Berlanga et al., 2011, [2]) or to use learning analytics as audience response technology or clickers in a blended environment (Mostyn, Meade and Lymn, 2012, [17]).

Aims or Problems Analysed on Formative Feedback Research in Online and Blended Environments

Another issue which has been analysed is *the aims that the papers propose to achieve*. By aims we are referring to the objective, research questions or hypothesis of the papers that have been analysed. This issue is presented together with the analysis of the papers' findings given that both issues are strongly related.

Concerning the aims, Table 3 shows that 28.9% of the papers focused on testing a tool to give feedback. For example, Berlanga et al. (2011) [2] use a tool which allows gathering data to create a concept map assisting the teacher to give feedback. Another example is Crook et al. (2012) [3] who tested a video technology delivered via the ASSET resource (this resource aims to explore the use of video as a means of enhancing the feedback experience for both students and staff).

Table 3. Aims and findings of formative feedback in online and blended environments across the 24 empirical studies included in this review

Aims and findings			
Aims	**Number of the papers (*see Table 1 to identify the papers*)**	**Frequency of papers (N = 24)**	**% of papers**
Testing a tool	2, 4, 3, 5, 9	5	28.9%
Gathering perceptions of students about feedback	1, 2, 3, 4, 6, 8, 12, 13, 15, 17, 19, 24	12	50%
Comparing two conditions	7, 10, 11, 14, 16, 18, 20, 21, 22, 23, 24	11	45.8%
Analysing the effects on learning	1, 6, 17, 24	4	16.7%
Findings			
Advantages and/or disadvantages of the tool or medium to give feedback	2, 3, 4, 5, 9, 11, 12, 13, 15, 19, 24	11	45.8%
Improving performance when receiving feedback	1, 6, 8, 17	4	16.7%
Comparing personalized vs. collective feedback	7	1	4.2%
Comparing F2F vs. online	10, 16, 18, 20, 22	5	20.9%
Comparing audio vs. written feedback	14, 23	2	8.4%
Comparing two tools to deliver feedback	1	1	4, 2

The categories in each question are not exclusive; that is, one paper can include different ideas on the aims and findings. For this reason, in each question, the frequencies of each category do not total 24; and the percentages do not total 100.

Another group of papers (50%) are focused on gathering students' perceptions of feedback. These perceptions may be related to their preferences, their satisfaction or what kind of feedback they consider more useful. Lin and Yang (2011) [12] is a good example of a paper that centres on students' perceptions of integrating wiki and peer feedback in writing skills.

Less than half of the papers (45.8%), concentrate on comparing two different learning conditions. Gallien and Oomen-Early (2008) [7] is illustrative of this type of paper, as they compare collective versus tailored feedback, aiming "to determine if there were significant differences in students' academic performance, satisfaction, and perceived "connectedness" scores when the instructor provided collective versus tailored feedback" (pp. 468). Another example of this line of enquire is the paper by Wieling and Hofman (2010) [22], whose research question is "to what extent a blended learning configuration of face-to-face lectures, online on-demand video recordings of the face-to-face lectures and the offering of online quizzes with appropriate feedback has an additional positive impact on the performance of these students compared to the traditional face-to-face course approach?" (pp 992). Both examples present a research where two different uses of feedback are compared.

Finally, there is another group of papers, 16.7%, whose aims focus on the analysis of the effects of feedback on learning. In this case, learning is measured considering several indicators. For example, Espasa and Meneses (2010) [6] used scores and satisfaction; Mostyn, Meade and Lymn (2012) [17] used the exams results and Xu (2010) [24] used student engagement and achievement in hybrid special education courses.

Problems, research questions or aims determine findings of the research. When analysing findings, the results obtained show that 45.8% of the papers focus on the advantages and disadvantages of the tool used when giving feedback. Among these papers there are some whose aims were focused on testing a tool, as we previously mentioned. Following the examples presented above, Berlanga et al. (2011) [2] concluded that "an initial validation, which explored tutors' perceptions of the tool showed that tutors found the approach relevant, but its implementation in practice requires to consider teachers' practices, the tools already in use, as well as institutional policies" (pp 11). Crook et al. (2012) [3] indicated that "a pilot of the ASSET resource for generic feedback provision found that it was considered advantageous by staff and students. Moreover, the use of video was also shown to resolve many of the common problems of feedback in relation to quality and engagement of students" (pp. 386).

There are other papers (16.7%) which present results highlighting the relevance of feedback in the sense that they ascertain that feedback influences positively on learning. One example of this type of papers is Espasa and Meneses (2011) [6]. The authors reported that there is a statistical relationship between feedback and learning results (students' satisfaction and final grades). This allowed the authors to assert the relevance of feedback in favouring self-regulatory competences within online teaching and learning practices. Another example of a paper that reported the positive relationship between feedback and learning is Getzlaf et al. (2009) [8] who concluded that "effective instructor feedback has positive outcomes for the students. In addition, the results of this study confirm that students want to be able to use the feedback provided by instructors in two ways. First, and not surprisingly, students want to use feedback to make improvements in future assignments. Secondly, students are interested in applying the feedback to practical situations in their daily lives" (pp. 15)

Concerning the papers that compare different conditions of feedback, through our analysis we were able to establish that comparisons are mainly focused on contrasting: tailored versus collective feedback, online feedback with face-to-face feedback, audio versus written feedback and the use of two different tools to give feedback.

Analysing this result in more detail, only one paper (4.2%), Gallien and Oomen-Early (2008) [7], focus on comparing the conditions among personalized and collective feedback. It concludes that "students who received personalized feedback were more satisfied and performed academically better than students who received only collective feedback" (pp. 463).

Around 21% of the papers reviewed presented findings that compare feedback given in face-to-face environment and feedback given online. In this comparison, there is no agreement between the papers analysed. However there are similarities between some of the papers which make up this group. For example, two of the papers reviewed, Moradi and Karimpour (2012) [16] and Oskoz (2009) [18] concluded that there is no difference between face-to-face and online. Both papers focus on language learning.

Two other papers which compare face-to-face and online feedback are Hwang and Arbaugh (2009) [10] and Wieling and Hofman (2010) [22]. Both are contextualized in blended learning. Findings of these papers reported the benefits of online teaching and learning practices as a complement to face-to-face teaching and learning practices.

Finally, the last paper found which compares feedback given in face-to-face environment with feedback given online is Shroff and Deneen's (2011) [20]. The authors conclude that "online textual feedback was strongly supported compared to feedback in a human non-technology supported face-to-face setting" (pp. 87).

Another type of comparison is presented in two other papers (8.4%), Macgregor, Spiers and Taylor (2011) [14] and Shroff and Deneen (2011) [20], which focus on comparing audio and written feedback. In a general view, both papers reported the potentialities of audio-feedback compared to face-to-face feedback.

Finally, the paper by Van der Pol et al. (2008) [21] presents two studies where two different tools to give feedback are compared. Results obtained in this paper regarding the different tools that were used, showed that concrete suggestions of feedback were more often produced in the Annotation system than in the Blackboard discussion forum, the latter showing more evaluative forms of feedback (pp. 1804).

In general, the analysis of the 24 papers reviewed highlight the problems or aims that have been analysed in the literature reviewed about feedback. The aims or problems which prevail are those related with students' perceptions (50%), on the one hand, and the comparison between two different conditions of feedback (45.8%) on the other. Concerning the findings, 33.4% of the papers reviewed presented findings about the advantages and disadvantages of the feedback tools. Around 21% of the papers presented results on feedback given in face-to-face or online environments.

Methodological Approaches in Formative Feedback Research in Online and Blended Environments

In terms of methods, three issues have been analysed. On the one hand, data sources and data analysis; on the other, the sample.

Data sources refer to the instruments which have been used to gather the data. From the analysis of the papers selected, nine different instruments have been identified: surveys, interviews, self-reported reflections, texts (i.e., comments, post which is considered as feedback), online conversation, focus group, students' observation, and revision of documents and scores.

As Table 4 shows, almost 71% of the research presented in the papers has used surveys to gather data. That means that the majority of the papers have a quantitative methodological approach. It is also relevant that half of this research has used interviews. It's worth highlighting that some research, 29.2% to be precise, has used both, surveys and interviews. An example of the combination of both methodologies is the paper by Jones, Georghiades and Gunson (2012) [11] where interviews were first conducted in order to gather students' reactions to the feedback received and then a survey was administered which allowed the authors to generate quantifiable data.

Table 4. Methodology used to study formative feedback in online and blended environments across the 24 empirical studies included in this review

Methods			
Data source	**Number of the papers** (*see Table 1 to identify the papers*)	**Frequency of papers (N = 24)**	**% of papers**
Questionnaires/surveys	1, 2, 3, 4, 5, 6, 7, 8, 11, 10, 12, 14, 15, 17, 22, 23, 24	17	70.9%
Interviews	11, 19, 20, 5, 1, 12, 14, 17, 19, 20, 24	12	50%
Self-reported reflections	12	1	4.7%
Text (Piece of feedback, comments)	21, 24, 2, 1	4	16.7%
Online conversation/dialog	1, 4, 18	3	12.5%
Focus group	2, 11, 12, 13	4	16.7%
Students' observations	12, 20	2	8.4%
Documents	13	1	4.7%
Scores	16, 17, 22	3	12.5%
Do not specify	9	1	4.7%
Data Analysis			
Descriptive analysis	3, 11, 15	3	12.5%
Relationship among variables (statistics)	1, 6, 7, 10, 14, 16, 17, 22, 22, 24	10	41.7%
Thematic/content analysis	1, 2, 8, 4, 13, 14, 17, 18, 19, 20, 22, 23, 24	13	51.2%
Grounded theory	11	1	3.7%
Do not specific	5, 9, 12	3	11.1%
Sample*			
<50	2, 4, 8, 9x2*, 12, 14, 18, 19, 20, 22x2*, 24	13	51.2%
Between 50 and 100	1, 7, 11, 15, 16,	6	22.2%
Between 100 and 150	11, 5x2*	3	11.1%
Between 150 and 200	6	1	3.7%
Between 200 and 250	10	1	3.7%
>250	3, 5, 22	3	11.1%

The categories in each question are not exclusive; that is, one paper can include different ideas on the methodologies and can simultaneously introduce two o more different methodologies on feedback. For this reason, in each question, the frequencies of each category do not total 24; and the percentages do not total 100.

* In this case, N = number of studies in the articles, taken into account that three papers (nº5, nº9, nº22) present two studies in the same paper. Then N = 27.

Four papers (16.7%) used focus groups as a source of data. In general, focus groups have been used to identify the advantages and disadvantages of using a given tool, or using a specific methodology –peer-feedback- for giving feedback. In the case, for example, of Dysthe and Lillejord's paper (2012) [13], the focus group had the objective of gathering the opinions of students on central aspects when creating an online community.

Some other papers reviewed (16.7%), have analysed feedback comments or feedback posts. This kind of analysis is associated with a qualitative approach. Another group of papers (12.5%) has obtained data from online conversations. This means that the online interaction among several actors of the teaching and learning process (teacher-students or students among themselves) is analysed. This kind of data source is also linked to a qualitative approach and to a thematic data analysis. An example of this is Alvarez, Espasa and Guasch's (2009) paper [1]. The authors defined a unit of analysis based on episodes corresponding to a fragment of joint activity (segment of interactivity) made up of the teacher's intervention and the student's response to the feedback received. Each of these episodes was categorized through a content analysis.

Finally, there are three papers that based their research on scores. These score are a mean which, in each paper, has been calculated taking into account a specific criterion. For example, Moradi and Karimpour (2012) [16] scored the writing pieces. "As the scoring ran the risk of misjudgment, both the researcher and the class instructor rated the pieces so that the average score could emerge" (pp. 115). To conclude this section, it is worth mentioning that one paper used self-reported reflections; another paper used documents (i.e., institutional documents are analysed). (See Table 4).

With regards to the *data analysis*, four different types were identified: descriptive; based on the statistical relationship among several variables; thematic analysis and grounded theory. The papers performing a descriptive analysis are those presenting the results based on percentages. These are three papers (12.5%) of this type. Other papers present a statistical analysis based on relationship among variables (41.7%). There are papers which have analysed data with thematic or content analysis (51.2%) and finally there is one paper that has used the Grounded Theory (3.7%).

The papers using surveys/questionnaires as their data source had a quantitative methodological approach. In these cases, the data analysis was mainly based on the relationship among variables, and the findings were obtained from the comparison between two different conditions of feedback. Papers whose data source was interviews followed a qualitative methodological approach and their findings were largely focused on the advantages and disadvantages of the technology to deliver feedback.

The last issue to comment on in relation to the methods is *the sample*. Table 3 shows that 51.2% of the papers reviewed present a research whose sample is made up of less than 50 subjects. 22.2% of the papers had a sample of between 50 and 100 subjects. Only two papers (7.4%) have a sample of between 150 and 250 subjects. Three papers (11.1%) have a sample of more than 250 subjects. In general, a review of the papers selected show that research on formative feedback in online and blended environments is conducted with limited samples.

DISCUSSION

The aim of this chapter was to identify general trends on feedback research in online and blended environments in higher education during a recent period from 2008 to 2012. To discuss the results obtained in this chapter, we will follow the same rationale that has been followed until now. First of all, a discussion of theoretical approaches, feedback definition and the use of ICT to give feedback; secondly, a discussion of the problems and questions that emerged in the literature reviewed, and thirdly, the methodological approaches identified.

In relation to theoretical approaches, three issues are worth noting: 1) the analysis of the papers selected show that formative feedback has been studied from different theoretical perspectives (there is no consensus on this matter); 2) theoretical approaches on formative feedback in online and blended environments are more closely linked to a sociocultural perspective than to objectivism, as has traditionally been the case. The study of co-construction of knowledge (peer-feedback), the teacher support or the consideration of students' needs, are issues found in the review of the theoretical approaches of formative feedback, which are characteristics of the sociocultural perspective. Accordingly to Hattie and Gan (2011), a sociocultural framework could be useful for research on feedback processes; 3) some studies do not mention the theoretical approach underpinning their research, which leads us to infer that the research studies analysed focus on the results that may be achieved and not so much on the theoretical approaches to feedback.

In terms of the definition of feedback adopted in the various papers under review, we expected to find prevalence of the dialogic perspective of feedback. This perspective, as defined in the introduction to this chapter, places emphasis on the students' use of feedback and gives greater importance to the process than to the product of feedback. However, it was found that the papers reviewed do not actually follow this perspective but they rather focus more on the act of giving feedback instead of on the analysis of the usefulness of the feedback received, thus evidencing little presense of the dialogic functions of feedback.

In terms of feedback components, the research papers reviewed focus on the study of one or two of the three feedback components proposed by Narciss (2008): content, form and function. However, we were unable to find any multidimentional approaches grounded on an integrated analysis of feedback. Concerning the main feedback components, the papers reviewed show a greater presence of research focused on the structure or form in which feedback is shaped (they attempt to analyse how Technology influences the form feedback is given in an online or blended environment). Technologies in general are the means by which feedback is given and, therefore, they influence the feedback shape. With regards to the use of technologies, the results obtained can be classified following two main uses. On the one hand, the technology becomes a means by which feedback is given: video, tablet PC, etc. According to Hepplestone et al. (2011), results of research on formative feedback in online environments ascertain that the technology through which feedback is given can encourage better engagement of students with feedback. On the other hand, the technology becomes a tool, which enables the teacher to gather evidences on how students are progressing in their learning. In this case, technology becomes the support that allows teachers to obtain objective data to deliver feedback. This technology use could be close to the visibility learning and teaching perspective (Hattie and Gan, 2011), mentioned in the introduction of this chapter.

The second issue analysed is the main problems or research questions found in the research on feedback in online environments. When analysing formative feedback in this context, questions are focused to analyse the effects of the technology used to give feedback. In this regard, two trends seem to emerge: 1) gathering perceptions in order to identify advantages and disadvantages of using technology to deliver feedback; and 2) comparing feedback given in face-to-face contexts with feedback given in online contexts in order to learn which feedback students prefer, which is more useful or which influences students' learning. In general, it seems there is no agreement on the benefits of giving feedback through a specific tool to improve learning. It is dependent on the tool and on the learning situation itself.

Finally, in relation to the methods there is evidence of both types of approaches: qualitative and quantitative. Taking into account the four generations proposed by Redecker and Johannessen (2013), the results obtained seem to be placed between the third and fourth generation, which highlights the notion of embedded assessment into teaching and learning. In terms of methodology, these generations -the fourth generation more so than the third-, are characterised by the learning analytics. In the papers reviewed there is small evidence of the presence of this methodology; however the intention of gathering data in order to be able to give feedback more adjusted to the students' needs is evident in several papers. Another relevant issue related to the methods is the fact that the most frequently used instrument for the collection of data was the questionnaire, which offers information on user perception but not on the actual use of feedback they make. The last issue in regards to the methods is the small samples used in their research. This doesn't allow for extrapolation of the results into other learning situations, and thus their validity is limited to a specific context.

CONCLUSION

After reviewing the research on formative feedback conducted during 2008-2012 years, the common trends found are presented below:

- Theoretical frameworks are situated closer to socioculturalism than objectivism.
- The role of the technology is oriented in two directions: 1) delivering feedback through tools such as video, tablets PC or audio feedback as already in use nowadays 2) gathering evidences of the learning process in order to give the teacher more information for giving feedback adjusted to the students' needs and to reduce teacher workload.
- Research on feedback structure/form (i.e., means: audio, video… by which feedback is given) is becoming one of the central topics analysed. No multidimensional approaches to feedback were identified.
- There is hardly any evidence of a dialogic perspective of feedback; that is, the research papers reviewed focus more on giving feedback that in the actions that students do with the feedback received.

- The methodologies used are both qualitative and quantitative, mainly based on questionnaires, and the samples are small.

To conclude this chapter, we would like to discuss some aspects that may guide future lines of research on feedback processes. On one hand, future research on this topic should include other levels apart from Higher Education, especially Secondary Education where feedback also plays a significant role. When selecting papers, some references to Secondary Education were found, which we did not consider, as they did not meet our criteria. Another issue that should be considered for future research is the assignment (i.e., study cases, essays…) on which feedback is given. In this overview, the kind of assignment has not been taken into account and some authors have pointed out this is a relevant issue in feedback research because it can influence the kind of feedback delivered. Finally, there is one last issue that could be interesting for future research, which is the consideration of the disciplines where feedback is researched. With regards to this, a preliminary analysis has been done. It shows that feedback is being researched in several disciplines (economics, education, engineering, medicine for example) and it would be interesting to learn what tendencies are evident in each of the disciplines when facing the challenges posed by giving formative feedback via technology.

REFERENCES

Alvarez, I., Espasa, A. and Guasch, T. (2012). The value of feedback in improving collaborative writing assignments in an online learning environment. *Studies in Higher Education*, *37* (4), 387-400.

Berlanga, A. J., Kalz, M., Stoyanov, S., van Rosmalen, P., Smithies, A. and Braidman, I. (2011). Language technologies to support formative feedback. *Educational Technology and Society*, *14* (4), 11-20.

Boote, D. and Beile, P. (2005). Scholars Before Researchers: On the Centrality of the Dissertation Literature Review in Research Preparation. *Educational Researcher*, *34* (6), 3–15.

Booth, A., Papaioannou, D. and Sutton, A. (2012). *Systematic approach to successful literature review*. London: Sage.

Bruce, C. S. (2001). Interpreting the scope of their literature reviews: Significant differences in research students concerns. *New Library World*, *102*(4), 158–166.

Carless, D. (2015). Excellence in University Assessment: Learning from award-winning practice: London: Routledge.

Carless, D., Salter, D., Yang, M. and Lam, J. (2011). Developing sustainable feedback practices. *Studies in Higher Education*, *36*(4), 395-407.

Coll, C., Rochera, M. J. and De Gispert, I. (2014). Supporting online collaborative learning in small groups: Teacher feedback on learning content, academic task and social participation. *Computers and Education*, *75*, 53-64. DOI:10.1016/j.compedu.2014.01.015.

Crook, A., Mauchline, A., Maw, S., Lawson, C., Drinkwater, R., Lundqvist, K. and Orsmond, P. (2012). The use of video technology for providing feedback to students: Can it enhance the feedback experience for staff and students? *Computers and Education*, *58* (1), 386-396.

Davies, S. (2010). "Effective Assessment in a Digital Age A guide to technology-enhanced assessment and feedback", Joint Information Systems Committee (JISC), U. Available from: http://www.webarchive.org.uk/wayback/archive/20140614115719/http://www.jisc.ac.uk/media/documents/programmes/elearning/digiassass_eada.pdf (Retrieved Jannuary, 30, 2016).

Dekhinet, R. (2008). Online enhanced corrective feedback for ESL learners in Higher Education. Computer I. (2010). Enhancing the impact of formative feedback on student learning assisted *Language learning*, *21* (5), 409-425.

Dexter, S. (2010). E-feedback intersections and disconnections in the interests of designers and users. *International Journal of Enginnering Education and life-long leraning*, *20*(2), 169-188.

Dysthe, O. and Lillejord, S. (2012). From Humboldt to bologna: Using peer feedback to foster productive writing practices among online master students. *International Journal of Web Based Communities*, *8*(4), 471-485.

Espasa, A. and Meneses, J. (2010). Analysing feedback processes in an online teaching and learning environment: An exploratory study. *Higher Education*, *59*(3), 277-292.

Espasa, A., Guasch, T. and Alvarez, I. M. (2013). Analysis of feedback processes in online group interaction: a methodological model. *Digital Education Review*, *23*, 59-73. (Retrieved January, 30, 2016).

Gallien, T. and Oomen-Early, J. (2008). Personalized versus Collective Instructor Feedback in the Online Courseroom: Does Type of Feedback Affect Student Satisfaction, Academic Performance and Perceived Connectedness with the Instructor? *International Journal on E-Learning*, *7*(3), 463-476.

Getzlaf, B., Perry, B., Toffner, G., Lamarche, K. and Edwaeds, M. (2009). Effective instructor feedback: Perceptions of online graduate students. *Journal of Educations Online*, *6*(2), 1-22.

Gikandi, J. W., Morrow, D. and Davis, N. E. (2011). Online formative assessment in higher education: A review of the literature. *Computers and Education*, *57*, 2333–2351.

Hart, C. (1999). *Doing a literature review: Releasing the social science research imagination*. London: SAGE.

Hattie, J. and Gan, M. (2011). Instruction based on feedback. In Mayer, R.E. and Alexander, P. (eds). *Handbook of research on learning and instruction*. New York: Routledge. Taylor and Francis Group. pp. 249-271.

Hatziapostolou, T. and Paraskakis, I. (2010). Enhancing the impact of formative feedback on student learning through an online feedback System. *Electronic Journal of e-Learning, 8*(2), 111-122.

Hepplestone, S., Holden, G., Irwin, B., Parkin, H. P. and Thorpe, L. (2011). Using technology to encourage student engagement with feedback: a literature review. *Research in Learning Technology*, *19*(2), 117–127.

Hwang, A. and Arbaugh, J. B. (2009). Original article: Seeking feedback in blended learning: Competitive versus cooperative student attitudes and their links to learning outcome. *Journal of Computer Assisted Learning*, *25* (3), 280-293.

Jones, N., Georghiades, P. and Gunson, J. (2012). Student feedback via screen capture online video: Stimulating student's modified action. *Higher Education*, *64* (5), 593-607.

Lee, S. J., Srinivasan, S., Trail, T., Lewis, D. and Lopez, S. (2011). Examining the relationship among student perception of support, course satisfaction, and learning outcomes in online learning. *Internet and Higher Education*, *14*, 158-163.

Lin, W. C. and Yang, S. C. (2011). Exploring students' perceptions of integrating Wiki technology and peer feedback into English writing courses. *English Teaching*, *1*, (2), 88-103.

Macgregor, G., Spiers, A. and Taylor, Ch. (2011). Exploratory Evaluation of Audio Email Technology in Formative Assessment Feedback. *Research in Learning Technology*, *19* (1), 39-59.

McVey, M. (2008). Writing in an Online Environment: Student Views of “Inked” Feedback. International *Journal of Teaching and Learning in Higher Education*, *20*(1), 39-50.

Moradi, M. R. and Karimpour, Z. (2012). The effect of online peer feedback on the academic writing ability of iranian EFL learners. *International Education Studies*, *5*(2), 113-117.

Mostyn, A., Meade, O. and Lymn, J. S. (2012). Using audience response technology to provide formative feedback on pharmacology performance for non-medical prescribing students - A preliminary evaluation.BMC Medical Education 12 (1). Available from http://www.biomedcentral.com/1472-6920/12/113 (Retrieved Jannuary, 30, 2016).

Narciss, S. (2008). Feedback strategies for interactive learning tasks. In J.M. Spector, M.D. Merrill, J.J.G. van Merrienboer and M.P. Driscoll (Eds.), *Handbook of Research on Educational Communications and Technology* (3rd ed., pp. 125-144). Mahwah, NJ: Lawrence Erlbaum Associates.

Nicol, D. (2010). From monologue to dialogue: improving written feedback processes in mass higher education. *Assessment and Evaluation in Higher Education*, *35*(5), 501-517.

Oskoz, A. (2009). Learners' Feedback in Online Chats: What Does It Reveal about Students' Learning? *Calico Journal*, *27*(1), 48-68.

Parkin, H. J., Hepplestone, S., Holden, G., Irwin, B. and Thorpe, L. (2012). A role for technology in enhancing students' engagement with feedback. *Assessment and Evaluation in Higher Education*, *37*(8), 963-973.

Redecker, C. and Johannssen, Ø. (2013). Changing Assessment —Towards a New Assessment Paradigm Using ICT. *European Journal of Education*, *48*(1), 79-92.

Shroff, R. H. and Deneen, Ch. (2011). Assessing Online Textual Feedback to Support Student Intrinsic Motivation Using a Collaborative Text-Based Dialogue System: A Qualitative Study. *International Journal on E-Learning*, *10*(1), 87-104.

Van der Pol, J., Van den Berg, I., Admiraal, W. and Simons, R. J. (2008). The nature, reception, and use of online in higher education. *Computers and Education*, *51*(4), 1804-1817.

Wieling, M. B. and Hofman, W. H. A. (2010). The Impact of Online Video Lecture Recordings and Automated Feedback on Student Performance. *Computers and Education*, *54*(4) 992-998.

Wood, K. A., Moskovitz, C. and Valiga, T. M. (2011). Audio feedback for student writing in online nursing courses: Exploring student and instructor reactions. *Journal of Nursing Education*, *50*(9), 540-543.

Yang, M. and Carless, D. (2012). The feedback triangle and the enhancement of dialogic feedback processes. *Teaching in Higher Education*, DOI:10.1080/13562517.2012.719154 Available from http://dx.doi.org/10.1080/13562517.2012.719154 (Retrieved Jannuary, 30, 2016).

Xu, Y. (2010). Examining the effects of online feedback on student engagement and achievement. *Journal of Educational Computing Research*, *43*(3), 275-291.

INDEX

A

B

C

D

E

F

G

H

I

J

K

Q

R

S

T

U

V

W

Z